BACTERIA IN BRITAIN, 1880–1939

STUDIES FOR THE SOCIETY FOR THE SOCIAL HISTORY OF MEDICINE

Series Editors: David Cantor
Keir Waddington

TITLES IN THIS SERIES

FORTHCOMING TITLES

BACTERIA IN BRITAIN, 1880–1939

BY

Rosemary Wall

Routledge
Taylor & Francis Group

LONDON AND NEW YORK

First published 2013 by Pickering & Chatto (Publishers) Limited

Published 2016 by Routledge
2 Park Square, Milton Park, Abingdon, Oxfordshire OX14 4RN
711 Third Avenue, New York, NY 10017, USA

First issued in paperback 2015

Routledge is an imprint of the Taylor & Francis Group, an informa business

BRITISH LIBRARY CATALOGUING IN PUBLICATION DATA

Wall, Rosemary, author.
Bacteria in Britain, 1880–1939. – (Studies for the Society for the Social History
of Medicine)
1. Medical bacteriology – Great Britain – History – 19th century. 2. Medical
bacteriology – Great Britain – History – 20th century. 3. Public health – Great
Britain – History – 19th century. 4. Public health – Great Britain – History –
20th century. 5. Bacterial diseases – Great Britain – History – 19th century. 6.
Bacterial diseases – Great Britain – History – 20th century.
I. Title II. Series
616.9'201'0941-dc23

ISBN-13: 978-1-138-66230-8 (pbk)
ISBN-13: 978-1-8489-3427-6 (hbk)

Typeset by Pickering & Chatto (Publishers) Limited

CONTENTS

In memory of John Cresswell and Hilary Bodsworth

ACKNOWLEDGEMENTS

This book has developed from a PhD undertaken at Imperial College London and hence I have many people to thank for their help along the way. Firstly, thanks go to Andrew Mendelsohn, my PhD supervisor and later my colleague at Imperial College London. Andrew inspired me with his ambition and enthusiasm, particularly his advice that hospital case notes would provide a wealth of information for my project. As second supervisor, David Edgerton has been very supportive of my research and career. Abigail Woods offered advice on public perceptions of disease and the chapters which examine interactions between humans and animals. The research environment at Imperial College encouraged me to pursue my interests in an expansive way, leading me to compare the use of bacteriology in a range of locations and settings, and I would like to thank all of the staff and students with whom I studied and worked at the Centre for the History of Science, Technology and Medicine between 2003 and 2006, and from 2011 to 2012. Many thanks also to my new colleagues at the University of Hull, especially Richard Gorski and Joy Porter, for their supportive advice and for making me think about aspects of my work in a different way. I would also like to thank Sally Sheard and William Ashworth at the University of Liverpool for nurturing my interest in the history of medicine, science and bacteriology.

Thank you to the Arts and Humanities Research Council for funding the PhD (award number 103736) and to Imperial College and the University of Hull for supporting three research trips to Geneva. This book could not have been written without this research funding and the support of a range of archivists. I am very grateful for access to a variety of archives and libraries which I list in the order of time spent within them: St Bartholomew's Hospital, Addenbrooke's Hospital, Croydon Local Studies Library and Archives Service, the Royal College of Physicians, West Yorkshire Archives Service in Bradford, North Yorkshire Record Office, Gwynedd Archives at the Caernarfon Record Office, University of Bradford Special Collections, International Labour Organization, Marlborough College, Zoological Society of London, Whitgift School, Cambridgeshire Collection and the County Record Office in Cambridge, Lincolnshire Archives, and University of Wales, Bangor, Department of Archives and Manuscripts. I have used archives and libraries extensively and any errors or omissions in seeking permissions or in crediting thanks are unintentional. In particular, I am especially grateful for the support and friendship of Chris Bennett at the Croydon Archives Service and Hilary Ritchie at Addenbrooke's Hospital, and for my time spent at St Bartholomew's Hospital Archives where I was ably assisted by Samantha Farhall, Marion Rea and Katie Ormerod. Linda Turnbull at the North Yorkshire Record Office was very happy to share her local knowledge with me. I would also like to particularly thank the very helpful staff at the International Labour Organization and United Nations for facilitating my brief but fruitful visits to the archives in Geneva – I have more plans for another publication using the materials I found there. I have also been very lucky to be based in London for ten years with easy access to the excellent Wellcome Library. John and Joan Cresswell, Peter and Hilary Bodsworth, Kate and Jim Wall, and Sue Redfern very kindly hosted me during research trips. Kurt and Charlotte Weinberg's generosity provided an affordable home in Nansen Village during my PhD studies.

Earlier drafts of some of the content within Chapters 1 and 2 have appeared as articles. I am grateful to Oxford University Press for permission to republish material from 'Using Bacteriology in Elite Hospital Practice: London and Cambridge, 1880–1920', *Social History of Medicine*, 24 (2011), pp. 776–95. The editor of *Forum Qualitative Sozialforschung* has very kindly allowed me to republish revised sections from '"Natural", "Normal": Discourse and Practice at St. Bartholomew's Hospital, London, and Addenbrooke's Hospital, Cambridge, 1880–1920', *Forum Qualitative Sozialforschung*, 8 (2007), art. 17. Thanks also to the Croydon Local Studies Library and Archives Service for permission to reproduce Figures 1.1, 6.1 and 6.2.

I would also like to express my gratitude to the people who have read sections of my book or who have provided me with invaluable feedback and support. Andrew Mendelsohn, Martin Gorsky and Michael Worboy, my supervisor and examiners, read the whole PhD thesis and provided a wealth of ideas of how to improve and clarify my arguments. Subsequently, Gayle Davis, Andrew Mendelsohn and Abigail Woods have read draft chapters of the book and provided insightful advice on how to present my ideas. I am also very grateful to the anonymous Pickering & Chatto reviewer who mentored me through the process of turning my thesis into a book. During my time at King's College London, I had many fruitful discussions with Anne Marie Rafferty regarding the process of writing my first monograph. More broadly, I also benefited from discussions with colleagues with an interest in history in the Florence Nightingale School of Nursing and Midwifery, the interdisciplinary Centre for the Humanities and Health, and within the History of Medicine reading group which included Ludmilla Jordanova, Katherine Foxhall, Keren Hammerschlag, Florence Grant and Richard McKay, and which attracted more and more members during my time at the College. I have also received valuable feedback following many seminar and conference presentations and conversations about my work. I would particularly like to thank Steve Sturdy, Joseph Melling, John Harley Warner, Virginia Berridge and Kati Hannken-Illjes for new insights on how to further my research. My mother and husband have helped with their expertise in polishing the final manuscript: Joan Cresswell's close reading has improved the text, and Joseph Wall has not only read the whole book but has continually supported me with his medical knowledge, expertise with computers and his patience. The book could not have been completed without him.

Ornella Moscucci and Keir Waddington have been very supportive in encouraging me to submit my work to the Pickering & Chatto/SSHM series. In particular I would like to thank Keir and my editors, Ruth Ireland and Mark Pollard, and copy-editor Frances Lubbe, for their superb efficiency, patience and kindness, making the process of publishing my first book as painless as possible.

This book is written in memory of two inspiring relatives who provided kindness, support, wise advice and, most importantly, humour to many people during their lives: my father, John Cresswell, and my aunt, Hilary Bodsworth.

LIST OF FIGURES AND TABLES

INTRODUCTION

In 1937, Croydon, a town ten miles south of London, was gripped by a typhoid epidemic which killed forty-three people. One of the victims was Richard Rimington, a schoolboy aged only thirteen. As soon as Richard became ill, his father, Charles, rose to the challenge of discovering the cause of his son's disease and that of his neighbours. With the local medical profession apparently stupefied by the outbreak, Charles Rimington, his friends and the citizens of Croydon helped to unearth the means of transmission. The outbreak led to charges of miscommunication within local government, resulting in what was probably the first successful mass case for compensation for a British epidemic. The science of bacteriology was seen as the key to confirming the cause of the epidemic and, as illustrated in Figure I.1, citizens, lawyers and doctors were eager for the results of bacteriological tests.

Figure I.1: The chief administrator of the local council is depicted 'in the campaign against the epidemic' and a 'Croydon chemist' is illustrated testing the local water supply. *Daily Sketch*, 20 November 1937, Croydon Local Studies Library and Archives Service, Sir Walter Monckton, KC, 'Outbreak of Typhoid in Croydon, Nov. 1937, Press Cuttings November 4–25, Local Papers November 26–7, Volume 1, Town Clerk Croydon', fs70 (614.4) CRO, p. 61. Image reproduced courtesy of Croydon Local Studies Library and Archives Service.

This incident encapsulates the novel approach of this book to one of the modern world's most pervasive sciences: medical bacteriology.[1] The epidemic of bacterial disease in Croydon reveals a knowledgeable response from members of the local community and from lawyers. In each chapter of this book the esoteric technical field of medical bacteriology is examined through the eyes of people who were working in various arenas outside of the laboratory, whether in the wards, the workplace or in wider communities.

In 1880, when the book begins, proof of the bacterial aetiology of disease had only recently been announced and demonstrated. The story ends in 1939, just before the age of antibiotics, with its own history of experts and publics.[2] *Bacteria in Britain* is thus the first sustained study showing how the new science of bacteriology was not imposed upon, but used proactively and creatively by men and women in hospitals, workplaces and communities. A major contribution is the revision of the historiographical idea of the resistance to, or reluctant acceptance of, laboratory science by elite physicians, through detailed analysis of their practice recorded in nearly 2,000 hospital case notes. Another is that the book brings to light the use of bacteriological knowledge by lay people threatened with disease, and their collaboration with lawyers and doctors. By examining a variety of interweaving communities of doctors, employees, citizens and lawyers, the significant changes in beliefs, practices and the use of new technologies are illustrated. These changes were tempered by the continued use of existing methods of diagnosing, tracing and combatting disease in the hospital and in public health.[3]

The book is divided into three interconnected parts – the hospital, the workplace and the community. 'The Hospital' examines medical practice and policy, in particular the debates around the funding of places and people for bacteriological diagnosis, diagnostic tests ordered by physicians for the diseases pulmonary tuberculosis, diphtheria and typhoid, and the discourse and representations of hospital physicians with regard to laboratory science. The teaching hospital is a crucial venue for understanding the use of bacteriology as it is where influential leaders of medical practice consulted and where future doctors were trained. 'The Workplace' looks at anthrax in various settings, including the woollen and leather industries. It unearths the ideas of employees, the public and lawyers, and their interactions with physicians, surgeons and specialist bacteriologists. Industry provided an environment where employees and guilds could organize and complain about their unsafe conditions, utilizing the new science of bacteriology to justify their grievances. 'The Community' uses the history of typhoid in residential areas as a device to reveal and explore public knowledge of bacterial disease. How did citizens, lawyers and doctors use epidemiological methods and bacteriological test results to understand and respond to epidemics? The three parts of the book are linked by considering the impact of bacteriology on the authority of doctors, and how lay knowledge of this science could be used to challenge expertise.

Using Bacteriology

The wide variety of case studies enables *Bacteria in Britain* to present a new view of the history of medical bacteriology in Britain by exploring not the innovations but the use of the science and technologies of bacteriology.[4] The story is reconstructed from a large volume of unpublished primary sources and is informed by histories of everyday life and technology in use, providing an alternative to top-down historiography.[5] This approach is particularly appropriate for understanding the use of a body of knowledge which is applicable to the 'everyday practices of public health institutions, the clinic, and even one's bathroom'.[6] In order to explore the minutiae of various lives and environments, the key sources for this book are hospital case notes and minute books, personal papers, local newspapers, records of medical and workplace societies, correspondence and transcripts of an inquiry and a trial.

A particular goal of this book is to illustrate the extent to which knowledge of how to combat bacterial infections and to delineate blame and responsibility for bacterial disease was shared between doctors, lawyers and 'publics' in Britain.[7] Brian Wynne has argued that concepts of modernity and reflexive modernity have been too simplistic in dividing scientific and public expertise, and he seeks to problematize and blur the boundaries of expert and lay knowledge.[8] The 'publics' who practise 'citizen science' or 'popular epidemiology'[9] may be engaged with risk even before an incident or 'expert conflict' occurs, and, although their knowledge is often ignored, have contributed expertise in a variety of late twentieth-century contexts which are as diverse as the reduction of nuclear contamination of sheep and the study of menstruation.[10] However, as *Bacteria in Britain* shows, this form of citizen science is not novel. Patients, their families and those at risk have complained about disease and collaborated with and exchanged knowledge with doctors since at least the nineteenth century. Indeed, Jean-Baptiste Fressoz has argued that the 'risk society' should not be seen as a recent postmodernist condition, but historicizes it within the nineteenth century, arguing that complaints about risks led to 'safer technological systems'.[11] Keir Waddington and Abigail Woods have illustrated the importance of the role of the public in arguing for prevention of infectious disease in relation to cattle in the late nineteenth and early twentieth centuries.[12] This book examines public knowledge of the new science of bacteriology in this same period. It reveals that in the context of industrial disputes and arguments for communication between water engineers and public health doctors, members of the non-expert public and their legal representatives could be knowledgeable, influencing safer practices.[13] From the 1880s, doctors valued the opinions of 'lay experts' working in industry in Bradford where employees were at risk from anthrax, and in Croydon the citizens' local knowledge and contacts were very valuable for epidemiological investigations into the cause of the 1937 typhoid epidemic. Patients and at-risk groups have been seeking biomedical 'knowledge empowerment' and critiquing medical research and practice for a considerable time.[14]

Bacteriology in Britain

The scope of this book spans from 1880 to 1939, a period of great change and increasing complexity for bacteriological research and practice in Britain. Robert Koch's research in the late 1870s has been seen as pivotal in confirming that specific microbes caused human diseases. Koch developed reliable methods of growing and staining bacteria, and postulates for determining the pathogenic cause of a specific disease.[15] His research on infectious disease has been understood as the turn from theory into technology.[16] There was a rapid succession of discoveries including Koch's explanation of the life cycle of the anthrax bacillus in 1876, his discovery of the tubercle bacillus in 1882 and identification of the comma bacillus causing cholera in 1883. Other bacteriologists' discoveries include Carol Joseph Eberth's and Edwin Klebs' discovery of the typhoid bacillus in 1880–1, and Friedrich Loeffler's discovery of the diphtheria bacillus in 1883, to name but a few.[17] During Koch's career, bacteriology changed from an 'exotic body of knowledge to one of the leading disciplines of experimental medicine'.[18]

Developments in practices for diagnosis and prevention of typhoid can be used to illustrate the evolving techniques of bacteriology in the late nineteenth and early twentieth centuries. Hence, doctors, lawyers and the public had an ever-increasing range of concepts to grasp. Methods of defining typhoid became more and more complex, from the realization through anomalous serological diagnostic tests in 1896 that there were related paratyphoid organisms, to the classification of more strains of paratyphoids in 1918 and of typhoid from 1934.[19] Yet, new techniques were not uniformly used. For example, although the distinction between strains was argued to have confirmed the link between a carrier and the epidemic in the 1936 Bournemouth epidemic (see Chapter 5 of this volume),[20] this method was not useful during the course of the outbreak or at the inquiry or test case regarding the 1937 Croydon epidemic (see Chapter 6), and there was a continued and lengthy debate about the use of the anti-typhoid inoculation, discovered in 1898. The existence of bacteriophage – viruses that could consume bacteria – was suggested in 1915, but it was not until 1938 that extensive research into phage typing was published by James Craigie and Chun Hui Yen.[21] They collected 706 examples of strains from typhoid outbreaks in Canada, England and Scandinavia and tested them with phage; seventy-four examples were from England, including bacilli from the epidemics in Malton, Bournemouth and Croydon discussed in Chapters 5 and 6 of this volume. The scientists retrospectively linked carriers to the strains involved in the epidemics.[22]

This research resulted in the identification of ten different types of typhoid.[23] Craigie and Yen's phage research methodologies for typhoid were adapted to a range of other bacteria, including *Staphylococcus aureus*.[24] However, this technology was just too late to be of practical use for the epidemics studied in this book.

A major reason for the institutionalization of bacteriology and immunology in the late nineteenth and early twentieth centuries was this increasing technicality in methods. Gradmann proposes that Koch's key innovations were practical rather than theoretical. In establishing his postulates for bacterial aetiology in the 1870s and 1880s, Koch and his team created a set of techniques which involved staining, solid culture media, the Petrie dish, microphotography and animal experiments.[25] Edgar Crookshank's 1896 textbook on bacteriology literally illustrates the complexity of apparatus required for diagnosis and research by that date, from the microscope, media for preservation of fluids and tissues, cover-glass preparations, sterilization equipment, incubators and microphotographic apparatus, with over 100 pages of instructions regarding this equipment before any discussion of diagnosis begins.[26] Christopher Crenner has shown how some physicians in Boston, Massachusetts, ran laboratories from their private practices, but from the 1910s these laboratories became increasingly commercialized or institutionalized.[27] The expense and space needed for such apparatus meant that these activities could no longer easily take place in a physician's home or the side room of a laboratory.[28] As I discuss in Chapter 2, Bart's physician Thomas Horder wrote in his 1910 textbook on pathology that the information was intended to be useful for physicians in understanding how diagnosis was carried out, and not for them to undertake the practices.[29]

Bacteria in Britain contributes to the history of how British bacteriological laboratories were established for routine examinations. The detailed debates involved in funding and building laboratories are discussed, showing how physicians argued for hospital laboratories for diagnosis and research in London and Cambridge.[30] In hospitals and institutions before 1890, only King's College Hospital, the Royal Colleges of Physicians in Edinburgh and London, and the Brown Animal Sanatory Institution had experimental laboratories for bacteriological work. Researchers also worked on bacteriological experiments at St Thomas' and St Bartholomew's Hospitals. By 1894 other laboratories were opening, conducting practical teaching and occasional research, including those at St Bartholomew's Hospital, Guy's Hospital and University College London. Clinical laboratories undertaking bacteriological work were developed for local public health departments and hospitals in the 1880s and 1890s. Organizations such as the Royal Colleges of Physicians of Edinburgh and London, and the Clinical Research Association at Guy's Hospital offered diagnostic services for practitioners and hospitals from the 1890s, making specialist bacteriological services available to a wider range of physicians and general practitioners.[31]

At this time, the status and financial reward for pathology and bacteriology remained low, being considered 'service' work for bedside practitioners, as explored in Chapter 1.[32] Challenges were faced with regard to the inclusion of bacteriology in the medical curriculum, with the Royal College of Physicians of London deeming that there was no room for the subject as late as 1896. Extramural courses were run by some medical schools during the 1880s and courses were increasingly integrated into medical qualifications during the 1890s.[33] Vernon has argued that, in general, British bacteriology only became a speciality distinguished from public health and pathology in the interwar period when bacteriology commanded full-time posts. The discipline came of age by the Second World War, attracting philanthropic funding for research. Legislation such as the Notification of Disease Act (1889, compulsory from 1899) and the Venereal Disease Acts of 1915 had cemented the role of the laboratory as general practitioners' diagnoses were often confirmed bacteriologically by public health officials. The Wassermann reaction became a mandatory means of diagnosis for syphilis with the latter Act.[34]

Although the status of bacteriology was improving during the late nineteenth and early twentieth centuries, physicians were converting at different rates to the belief that bacteria were the cause of disease. Doubts about the new theory were partly as a result of demonstrations that healthy people secreted microbes, and that pathogenic microbes could be consumed and yet not result in disease. During the late nineteenth century many false discoveries of pathogenic bacteria were made which aided the discreditors of Louis Pasteur's and Koch's theories of disease.[35] Further to this, in 1890, Koch's tuberculin 'cure' for tuberculosis was dramatically announced and made commercially available within months. By the end of the year, its therapeutic value was doubted and it was found to be dangerous, with some patients dying following its administration.[36] Subsequently in Germany, an infamous challenge to the idea that infection with bacteria led to disease came from Max von Pettenkofer in 1892. In order to demonstrate his hypothesis that bacteria alone did not result in disease, he drank a concoction of cholera bacilli, resulting only in a little diarrhoea.[37] The puzzle as to why infection did not always result in disease, particularly for tuberculosis, led to continuity in ideas of the constitution of the body affecting disease susceptibility and to the discipline of immunology.[38] There was also a lack of consensus on the principle of specificity – of certain germs causing certain diseases – with ideas of transmutation of bacteria from one species to another in the 1880s and 1890s. Additional criticism included attacks on the technique of bacteriology for diagnosis of diphtheria with difficulties in finding the bacilli in cases which were clinically considered to be diphtheria.[39] Therefore, the introduction of bacteriological diagnosis in the hospital and into public health methodologies in the late nineteenth century by no means took place at a time when the precise nature and role of bacteria and bacteriology were completely defined and accepted.

Indeed, the authority of bacteriologists was questionable throughout the period of this study, with only a few successful therapies emerging for bacterial diseases. The therapeutic vaccine for rabies (1885), antitoxins for diphtheria, tetanus, anthrax and a range of other diseases (from the early 1890s), immunological vaccines, including for typhoid (1898), and Salvarsan as a treatment for syphilis (1909–10), brought hope that bacteriology would bear more fruit in terms of prevention and cure. These developments were slow to come, with chemotherapies which acted upon a wider range of diseases being developed from the 1930s.[40]

Yet, the practices created by early bacteriologists alerted doctors, employers, employees and the general public to the dangers to which they were exposed by organisms so miniscule that specialists with technical expertise and equipment were increasingly needed to reveal their characteristics. However, many of the practices revealed within the book hark from the era before bacteriology: boiling of milk and water with the aim of purification was not a new idea for example, and personal hygiene, the soap industry, private bathrooms and household cleanliness were developing in any case. Chapters 5 and 6 pay particular attention to the perceived dangers within the home which changed with publicity of the new science of bacteriology such as pasteurization of milk, representations of flies, views on the chlorination of water, and the decline of the concept of sewer gas. The three sections of this book could be framed as presenting the use of bacteriological resources and institutions within three sites – the increasingly routine use of the technologies of the laboratory in the hospital; the knowledge of aetiology of disease in the workplace; and the combination of bacteriological ideas and practices, such as the search for typhoid carriers, with older methodologies of epidemiology, in the community.

The Hospital

Providing concrete evidence for the assimilation of bacteriology in hospitals, one of this book's key contributions to historiographical debate is an analysis of 1,823 clinical case notes, in order to examine the everyday use of bacteriological diagnostic technology in the hospital.[41] Hospital committee minute books are also invaluable for discovering the everyday decision making and problem solving which occurs in the running of a hospital. Substantial evidence and analysis provide a revisionist approach to the history of the laboratory and clinic relationship, in particular the idea that gentlemen physicians were reluctant to incorporate the laboratory into their everyday practice.[42] Historiographical ideas of the tensions between the clinic and the laboratory led Andrew Cunningham and Perry Williams to conclude that hospital physicians were the most 'powerful ... group of those sceptical of the necessity, the usefulness or even the relevance of the laboratories' and that these physicians maintained 'fierce opposition in the early twentieth century'.[43] My research significantly furthers the small number

of studies which have used case notes to examine hospital use of bacteriological diagnosis in the first decades of the twentieth century. These existing studies
present a fairly enthusiastic reception of bacteriology by clinicians, though with
caveats that clinical decisions could override evidence from the laboratory, given
the possibility of false negative results.[44] Other publications revising this historiography have shown that more nuanced accounts are necessary in order to reveal
positive responses to the clinical relevance of laboratory medicine.[45]

Chapter 1 compares the funding of the laboratories and the increasingly
routine specialist diagnostic practices at St Bartholomew's Hospital and Addenbrooke's Hospital.[46] Pulmonary tuberculosis, typhoid and diphtheria were
chosen for study not only due to their prevalence in these two general hospitals
but also because of early identification of bacteria and diagnostic methods for
these diseases, and a successful immunological therapy for diphtheria. Given
that Brian Abel-Smith argued that general voluntary hospitals began to reject
cases of infectious disease from the 1860s, two large general hospitals may seem a
strange choice of focus for a study of infectious bacterial diseases.[47] The number
of typhoid cases at Addenbrooke's and Bart's, and of diphtheria cases at Bart's,
shows that, at least for these hospitals, patients with infectious disease were
commonly admitted. In 1907, the Medical Officer of Health for Cambridge
complained in his report that even though an Isolation Hospital had been constructed with a special ward for typhoid fever, fourteen cases of typhoid were
admitted to Addenbrooke's Hospital, in comparison to only one at the Isolation
Hospital.[48] At Bart's between twenty-one and forty-four cases of diphtheria were
admitted each year until at least 1920.[49]

In addition to the availability of extensive case notes, Bart's and Addenbrooke's have been chosen because of their heritage and the historiographical
representations of the hospitals. Although records for a variety of London hospitals were surveyed, the records at St Bartholomew's Hospital Archives captured
my attention. I quickly discovered that current representations of the use of
the laboratory and specialists were in need of revision as they were too reliant
on published sources, and that the language used in the case notes was very
different to that in American hospitals.[50] Bart's, the oldest continually open
hospital in Britain, is contrasted with Addenbrooke's, which was established
in the eighteenth century and quickly became used as the teaching hospital for
the University of Cambridge. The university was famous for laboratory science
during the period covered by this book, and the links between the hospital and
the university are considered, furthering Gerald Geison's and Mark Weatherall's
detailed studies of science and medicine in the town.[51] Despite recent historiography, the quick acceptance of bacteriological diagnosis is not really surprising
as teaching hospitals were encouraged to become more 'clinical and scientific'
due to the pressure of both licensing bodies and students in the late eighteenth

and early nineteenth centuries, and from the University of London from the mid-nineteenth century.[52] This trend continued in the 1890s with the General Medical Council extending the length of study for a degree in medicine from four to five years in 1892, in order to allow more time for science. Competition with provincial medical schools, which offered a higher standard of science teaching, also encouraged the development of laboratory facilities in London.[53]

Chapter 2 aims to explore how the integration of the laboratory into the practice of the gentleman physician was by no means straightforward. The chapter begins with an analysis of discourse in the case notes, revealing that the change in language from using the term 'natural' to using the term 'normal' when discussing the body was much slower at St Bartholomew's Hospital than at Addenbrooke's Hospital. The word 'normal' signified the standardization of measurements of the body.[54] The continued use of the word 'natural' at Bart's is intriguing considering the decline of its use in the American hospitals studied by John Harley Warner.[55] This examination of discourse is followed by two biographical studies of Bart's physicians Samuel Gee and Thomas Horder. Their individual use of diagnostic laboratory services is analysed in order to understand contradictions in public representations of their opinions on bacteriology, in contrast to their everyday use of specialist pathologists for diagnoses of their patients' conditions. The chapter links to the following sections on public knowledge with a discussion of Horder's opinions of his patients' views of the value of specialist diagnostic services in comparison with the more expensive expertise of an elite consulting physician.

The Workplace

Late-nineteenth-century anthrax provides an unrivalled opportunity to examine how ideas of bacteriology were received at the outset of publicity about the new science. The incidence of the disease was increasing in Bradford just as Koch's discovery of the lifecycle of the bacillus was publicized in the late 1870s. Although Koch's anthrax studies did not have a resounding impact on German medical researchers,[56] his discovery did influence workplace politics in Britain.

Anthrax was a risk to those who worked with animals and their skins and wool, and was occasionally a risk for the public, for example when shaving brushes were made from contaminated pony hair. Control of anthrax led to local, national and international investigations in order to understand how to combat the threat of the disease from abroad. Legal cases give the opportunity to further explore the early use of bacteriology in the courtroom, a topic which has been briefly explored by Waddington in relation to lawyers' use of new and uncertain evidence regarding the dangers of tuberculous meat in 1889.[57]

Bradford was highly politicized and was continually exposed to new revelations in laboratory research through publicity regarding five children and three men who had been bitten by a rabid dog early in 1886. They visited Pasteur's laboratory for the therapeutic rabies inoculation, only five months after Pasteur discovered the treatment in October 1885. The children became national celebrities.[58] Therefore, the study of Bradford in Chapter 3 is balanced with an examination of anthrax within the hide and skin trade in London in Chapter 4, where casual dockworkers were less organized but were able to enlist the support of doctors at Guy's Hospital.

As important in terms of morbidity as anthrax in Bradford, serious study of the disease in late nineteenth- and early twentieth-century Bermondsey, London, has been neglected by historians.[59] These cases also received less contemporary publicity. Textiles was a key trade for the British economy and woolsorters were well paid, whereas casually employed London labourers working with hide and skin were not always respected by their employers. However, by 1925 anthrax in leather was a high profile risk internationally. This problem became the first business of the Advisory Committee on Industrial Hygiene appointed by the International Labour Office of the League of Nations in 1923, and in which Britain reluctantly took a leading role.[60] Providing another London workplace comparison, an incident at Regent's Park Zoo, where four workers contracted anthrax, is examined to explore the risk of autopsying and disposing of diseased animals. Lastly, the publicity regarding anthrax caught from shaving brushes in the 1910s and 1930s by civilians and soldiers, resulting in a ban on brushes from Japan, will be analysed in order to compare reactions to risk of the disease which were not connected with occupational health, linking with the next chapters on public responses to bacterial disease.

The Community

Anyone could be at risk from typhoid, from the lower classes to Prince Albert, who died from the disease in 1861. The bacilli could infect a town's water supply, milk and also food. The carrier concept meant that typhoid had to be fought with both 'inclusive' and 'exclusive' measures, with bacteriology highlighting the need to concentrate efforts on the exclusive measures related to the typhoid sufferer or carrier – disinfection of bodily discharges – but also older sanitarian inclusive measures of tackling modes of transmission, such as monitoring the water supply and food.[61] Chapters 5 and 6 illustrate that although the practices of bacteriology were becoming increasingly complex, the basic concepts of bacteriology could be understood by the public, from the value of diagnostic tests to the role of carriers, and the hygienic practices which were required to prevent the spread of bacteria. A variety of case studies across England and Wales show gradual changes in public responses during this period, in tactics for protection from disease, to discovering and blaming those responsible. Chapter 5 explores the history of typhoid through a series of major epidemics in England and Wales between 1882 and 1936, particularly examining medical and public knowledge of the role of bacteriology in confirming epidemiological findings, the development of the carrier concept, and ideas regarding typhoid vaccination. The pivotal point of the chapter is a case study of typhoid in Malton, near York, in 1932, an epidemic during the economic depression which generated a wealth of correspondence.

The study of the 1937 typhoid outbreak in Croydon in Chapter 6 analyses a landmark medical and legal case, demonstrating how the father of a typhoid patient discovered the cause of transmission of the disease more quickly than the Medical Officer of Health, and brought together members of the local government in order to try to tackle the disease. The epidemic was followed by 260 claims for compensation. As John Fabian Witt has shown, there is no novelty in tort actions regarding death by negligence. In Britain, an Act authorizing actions was instituted in 1846, and American states quickly followed suit.[62] A typhoid epidemic in New York in 1928 resulted in a claim for $425,000 following 248 cases and 25 deaths, and therefore Britain was behind America in terms of successful mass litigation in court.[63] However, the 1897 Maidstone case, explored in Chapter 5, suggests that legal ideas of blame and responsibility confirmed by bacteriology began much earlier, at the same time as compensation began to be seriously discussed in relation to bacteria in industry.

Bacteria in Britain concludes by asking whether recent historians of the laboratory and clinic have been influenced by ideas of gentlemanliness, nostalgia and declinism. This last chapter situates the history of medicine within wider perspectives on the historiography of modern Britain. I argue that the new knowledge and technologies of bacteriology were quickly absorbed in late nineteenth- and early twentieth-century Britain by medical practitioners, lawyers, employees and the public. This led to fast integration of diagnostic practices in hospitals, increasing laboratory spaces in which specialists worked, and legal attempts to prove responsibility for disease.

1 USING BACTERIOLOGY IN TEACHING HOSPITALS: LONDON AND CAMBRIDGE, 1880–1920

[House Physician] – This is Mr. Cornwall's case.
Physician – Come here, Cornwall; what do you think is the matter with this patient?
Student – I should think she has phthisis.
Phys. – But stay! Cornwall, have you found the tubercle bacillus?
Stud. – No, but I expect to.
Phys. – We can hardly base our diagnosis on your expectations.
Stud. – No, but we will on her expectorations.

St. Bartholomew's Hospital Journal (1895–6)[1]

Historians have portrayed a tense relationship between physicians and surgeons diagnosing at the bedside and pathologists testing samples at the bench.[2] Yet, the unpublished records of two English hospitals – St Bartholomew's Hospital, London, and Addenbrooke's Hospital, Cambridge – tell a story of elite physicians who were enthusiastic about new diagnostic techniques carried out by specialist pathologists. Indeed, the comical epigraph for this chapter was published in Bart's hospital journal and represents just how routinely bacteriology was used for diagnosis at the hospital.

In order to understand the contradictory representations of the interactions between elite physicians in the clinic and the laboratory, this chapter analyses a variety of sources created in the course of everyday administration and care for patients within two hospitals. Use of bacteriological diagnosis at Bart's, the oldest continually open hospital in England, established in 1123, is compared with Addenbrooke's Hospital, which was opened to the public in 1766 as a voluntary hospital funded by a legacy from Dr John Addenbrooke, and increasingly developed links to the University of Cambridge.

Beginning with an analysis of representations of Bart's and Addenbrooke's, this chapter proposes that historians' views of the clinic and laboratory require revision, particularly when unpublished everyday hospital records are taken into account. First, financial and administrative debates are compared in order to reveal contrasting stories of how laboratories and specialists for bacteriological diagnosis were established at the hospitals. Second, the chapter examines the construction of hospital case notes at Bart's and Addenbrooke's, followed by an analysis of the use of diagnostic methods in 1,823 case notes which record the treatment of patients with pulmonary tuberculosis, diphtheria and typhoid.

St Bartholomew's Hospital

Medical education at Bart's emerged in the seventeenth or eighteenth century. By the 1880s, Bart's hosted the largest number of medical students in London and was perceived to be the most prestigious medical school in the city.[3] In his series of papers on London physicians, Christopher Lawrence has focused on physicians working at Bart's, with a major theme being the idea of gentility amongst these elite practitioners.[4] He attributes this characteristic to the 'distinguished lay governors' who selected physicians by 'class attributes rather than medical skill': the importance of 'character more than the pursuit of expertise'.[5] The perception that London's hospital physicians were different from those in the rest of the country is shared by other historians. Roy Porter describes them as holding 'plush positions of professional authority and eminence'.[6] These physicians controlled the professional institutions, holding the highest offices of the Royal Colleges of Physicians and Surgeons.[7] Jeanne Peterson best sums up the status of these physicians:

> Those who practiced in London hospitals and in private practice were rewarded with high social connections, prosperity if not wealth, and ... [the] pleasures of the upper ranks of Victorian society. Knighthoods or baronetcies, country houses, lavish entertaining, foreign travel, Alpine mountain-climbing, membership at the Athenaeum, art collecting – all were possible for the select few who were numbered among the London medical and surgical elite. They often married well, sent their sons to public school and university; they owned carriages, hired servants, and donned the morning coats and top hats that daily bespoke their elevated status.[8]

As recently as the 1960s, Paul Ferris, a prominent journalist, novelist and social commentator, claimed that the London teaching hospitals were 'still the seats of power', especially St Bartholomew's Hospital and St Thomas's Hospital, which had the 'best social standing and the nicest upper-class nurses'.[9] Ferris perceived that the London teaching hospitals remained 'stiff with protocol, old-fashioned and due to lose their grip on the central mysteries of medicine', with the London practitioner 'obsessed with the gentlemanliness of his profession'. He argued that the 'London system is still rooted in this archaic individualism of the part-time gentleman'.[10]

Gentlemen physicians have been portrayed as believing that clinical medicine was 'based on science' but, more importantly, that it was an 'art which necessitated that its practitioners be the most cultured of men and the most experienced reflectors on the human condition'.[11] Lawrence has examined the delayed use of medical instruments such as the sphygmograph and sphygmoma-nometer, as described in published medical literature.[12] For the interwar period, Lawrence argues that the gentlemen physicians approved of the laboratory for diagnostic and experimental purposes, but 'were determined to integrate it into their preferred social order' – an order which opposed specialization, mechani-zation and standardization, due to their holistic perspective.[13] In contrast to this account, histories of Bart's written in 1918, 1923 and 1961 celebrated the use of the laboratory and specialization by the hospital staff, whereas those written since 1974 portray a delay in these movements due to conservatism at Bart's, at least during the 1880s.[14] Chapters 1 and 2 and the conclusion discuss why there have been such conflicting representations of practice at the hospital. Why have representations of the history of Bart's changed so much over the twentieth cen-tury and what do hospital case notes and minute books reveal about everyday practices at the hospital?

Addenbrooke's Hospital, Cambridge

As a result of its relationship with the University of Cambridge, the introduc-tion of laboratory science at Addenbrooke's Hospital was more complex than at Bart's. By the 1880s, the university's school of medicine was developing a reputation to rival the London medical schools. Yet, the status of the school had been very low in the early nineteenth century. From 1833 to 1858, an average of less than four students per year studied medicine at Cambridge University, and the qualification obtained was not highly regarded by the medical profession.[15] Cambridge was compared unfavourably with London schools in medical jour-nals, particularly with regard to the size of the hospitals and a lack of dissecting rooms.[16] During the late nineteenth century this reputation changed, with sev-enty students beginning their medical studies in 1880. By 1891 the Cambridge medical school was one of the largest in the country, with the degree regarded as 'one of the highest professional qualifications'.[17] However, Addenbrooke's Hos-pital continued to be too small to accommodate all of the medical students for their whole degree, so London hospitals, amongst others, were used for student placements during the late nineteenth century.[18] In fact, this still happens today.

The Cambridge medical school aimed to attract physicians who had 'clinical ability and scientific curiosity' – people like John Bradbury, who was a physi-cian at Addenbrooke's for fifty years and who also published on pathological and clinical subjects in the *British Medical Journal* (hereafter *BMJ*) and the *Lancet*.[19] Indeed, Lawrence contrasts Addenbrooke's physicians with those in London, discussing Bradbury's and Clifford Allbutt's enthusiasm for 'scientific' clinical medicine, continental experimental science and their use of 'instruments of pre-cision', linked with Michael Foster and his Cambridge school of physiology.[20]

There were practical links between the hospital and the university. Addenbrooke's X-rays were initially carried out at the nearby Cavendish Laboratory, by members of the laboratory's staff, Mr W. H. Hayles and Mr Everett. These diagnostic X-ray images were produced as early as 1896, one year after they were discovered by Wilhelm Roentgen, and Hayles was probably the 'first man in England to take an x-ray'.[21] Addenbrooke's purchased its own machine in 1903. At first, surgeons unsuccessfully took their own photographs. In 1907, Mr Field, the dispenser, learnt how to use it from staff at Bart's.[22] Considering the historiographical representation of conservatism at Bart's, it is revealing that the dispenser at Addenbrooke's learnt about technology from colleagues at Bart's.

The university also had some political control over pathology at the hospital. In 1878, George Paget, Regius Professor of Physic, complained about the lack of consultation in the proposed election of a professor of physiology at Trinity College and how this typified the importance given to natural sciences compared to medicine at the university.[23] This hierarchy was made even clearer in 1883–4, when residents were given more power in decision-making than non-residents of the university. Therefore the Board of Medicine came to have a clear majority of pre-clinical scientists. This resulted in the first professor of pathology, Charles Smart Roy, being a physiological pathologist, rather than a clinical pathologist. The vote was actually tied between Roy and a clinical pathologist, David James Hamilton, but the Vice Chancellor of the University showed where his allegiance lay, choosing Roy.[24] At the news of his appointment, the *Lancet* announced that this would 'blight the hopes of those who were looking forward to the establishment of a practical medicine school at Cambridge', for Roy was really an experimental physiologist.[25]

Relations between the university and the hospital were not always amicable. The connection between the University of Cambridge and Addenbrooke's Hospital appeared assured as, from 1785, it had been the tradition that one of the physicians at Addenbrooke's Hospital was appointed the Regius Professor of Physic at Cambridge.[26] Other physicians at the hospital held additional posts which included Downing Professorships of Medicine, fellowships of various colleges and one Lucasian Professorship of Mathematics.[27] However, when Allbutt was appointed as the Regius Professor in 1892, this association came into question. Allbutt was appointed from Leeds, as it was too hard to choose between the three physicians at Addenbrooke's, and because Foster had a key part to play in the appointment. Indeed, Allbutt believed in the value of experimental science for the practice of medicine.[28] The Addenbrooke's physicians were angry about losing control of medicine within the university and perhaps contributed to the delay in Allbutt being accepted at the hospital.[29] He did not receive a position at the hospital until 1900, although he deputized for Donald MacAlister in the summers of 1896–7. In 1895, the Vice Chancellor asked the hospital for a conference to establish 'a closer and more regular connexion between the Hospital and the University Teachers in the Department of Medicine'.[30] Even though all medical and surgical members of staff at the hospital, bar one, were also professors or teachers at the university, and clinical lectures were given in the wards, the boardroom and the operating theatre, there was still no official link. In 1896, the committee agreed that lecturers could use the hospital and its patients for teaching purposes, but not alter patients' treatment. In 1900, an agreement was reached that the university would pay the hospital £300 per year, and the Regius Professor could be a physician, choosing whether he would like to supervise beds, or whether he would only like to use them for teaching purposes. Allbutt declined the offer of supervision, informing the hospital that he already had enough for his teaching requirements. He did not take over the care of any beds until the hospital was short of staff during the First World War.[31]

The University of Cambridge can be contrasted with Bart's because of the university's increasing reputation for laboratory science, and its links with Addenbrooke's Hospital. Professors were taking their jobs more seriously, considering an academic career could be for life, and were both transmitting and producing knowledge.[32] Foster's physiological school, the Cavendish Laboratory for physics, chemistry laboratories and anatomy museums were all changing the scientific culture of the university.[33] Foster, George Humphry, the Addenbrooke's surgeon, and George Paget, the physician, were the 'Great Triumvirate'. Together they influenced the rise of the successful medical school in Cambridge in the late nineteenth century.[34] Therefore, it is essential to consider the general scientific milieu in Cambridge in the late nineteenth and early twentieth centuries when examining changing practice.

Establishing the Laboratories

Although nineteenth-century Bart's physicians have recently been given negative press with regard to their adoption of technology, specialists working in the laboratory at Bart's have been presented as amongst those at the forefront of bacteriology in Britain. Emanuel Klein's bacteriology school crossed the various institutions in which he worked and was the most important in Britain, influencing public health, surgery and medicine. He worked at the Brown Animal Sanatory Institution, the College of State Medicine, and from 1873 was the lecturer in histology at Bart's.[35] Alfredo Kanthack was appointed as the lecturer in pathology at Bart's in 1893. Kanthack emphasized 'the importance of the daily use of the laboratory in medicine and surgery'.[36] The requirements of the Medical College were instrumental in this appointment, and the subsequent increased interest in laboratory diagnosis at the hospital. Bart's was trying to maintain its position as the largest medical school in the capital, but the Medical

College had been falling behind in recruitment and in its uptake of experimental and laboratory science. After Kanthack's appointment in 1893, student numbers rose for a while and, in 1894, a subcommittee of the Medical College appealed for a full-time hospital pathologist.[37]

The debate regarding the establishment of the post of hospital pathologist shows the hospital-wide support for specialization of bacteriological diagnosis. At a meeting between the Medical Council, the treasurer and the almoners in January 1895, the specific topic of discussion was 'the Pathologist'. The physicians at the hospital were represented by William Church and Norman Moore. The main spokesman for the post was Thomas Shore, lecturer in biology and comparative anatomy. However, Church appointed Shore to speak on the matter, demonstrating his support.[38] Although Church was clearly an eminent gentleman physician, he believed in the clinical value of the laboratory and specialists. He was educated at Harrow, Oxford and Bart's, was Reader of Anatomy at Christchurch College, Oxford, and was the first President of the Royal Society of Medicine.[39] Yet his case notes show that patients diagnosed with pulmonary tuberculosis, typhoid and diphtheria were tested using bacteriological methods in all but one instance after this became possible.[40] Church perceived original research to be part of the inevitable everyday work of the pathologist, and argued with the treasurer when the governors would not provide the money for this. The Medical Council wanted the pathologist to be paid £600, even though, as the treasurer pointed out, the highest salary for anyone on the medical staff was 100 guineas (£105). The medical staff were willing to contribute £2,000 to the fund, amounting to £60 a year, with the proviso that the governors would provide £10,000. The surgeon, Alfred Willett, wanted to call a public meeting in order to raise money. He proposed that the President, the Prince of Wales, should attend, as well as the governors and other 'influential residents in London'.[41]

In the previous year, the Medical Council had been advised not to campaign for a pathologist on the grounds that the 'public did not realise the importance of bacteriological investigation'. However, Shore felt the time was now ripe because of a new therapeutic discovery: 'The public can hardly fail to have a greater interest now than formerly owing to the great interest taken in the anti-toxin treatment of Diphth.'[42] Even so, it was still difficult to raise the money. The treasurer, Sir James Lawrence, understood their request, having been educated at Bart's, but he argued that even if the public recognized the worth of employing a pathologist, they apparently viewed Bart's as 'prodigiously and enormously rich' and did not donate money. It was hard to convince the governors too as they considered bacteriological research to be a 'scientific fad'. The treasurer considered the Medical Council's strongest claim for the appointment of a pathologist to be Shore's argument for necessity based on practical diagnosis rather than for scientific research: 'Med Off. have found during the past year a very great need of someone to whom they can officially refer cases in connection with the wards requiring further investigation beyond the ordinary methods of diagnosis, that is bacteriological investigation.'[43]

The doctors' mission to press for funding for a pathologist was partially successful. Kanthack's successor, Frederick Andrewes, claimed that Kanthack was the first clinical pathologist to be appointed in a London hospital.[44] Yet Kanthack was only paid £100 per year for his role of pathologist to the hospital, in addition to the sum that the medical school paid him for lecturing. Additionally, the governors did not provide the sum of £10,000 which was requested by the doctors.[45] Kanthack left to become professor of pathology at Cambridge in 1897.[46] In addition to the appointment of a pathologist, the clinical team were subsequently provided with laboratory provisions on the wards so that they could carry out some 'elementary' investigations. Church argued that a 'great amount of work' remained for the pathologist. In 1902, the Medical Council repeated its request for a larger salary for the pathologist, who was still only paid £100 a year, with his assistants working unpaid. This time the physicians Church and Wilmot Herringham represented the Council, and Howard Tooth, another physician, sent a letter.[47] Tooth's letter, and the meeting as a whole, showed the continuing commitment of the medical staff:

> Pathology and Bacteriology have been assuming an ever increasing importance in the investigation and treatment of the diseases of patients taken into the wards of the Hospital, and that it is on these lines that progress in diagnosis and treatment is to be expected in the future.[48]

Unlike members of the medical staff, Andrewes could not undertake private practice as his job entailed very long hours at the hospital. Church complained that his 'present payment' did not reflect the 'importance' of his work which was 'absolutely necessary to the proper treatment of patients', and that Kanthack had already been lost due to the poor salary. The treasurer argued that there were other reasons for Kanthack's resignation, and invoked the professional requirements of the physician, indicating that diagnosis was in any case an important part of the physician's role: 'unless the Physician were ready to undertake the work his services would diminish in value.'[49] However, the surgeon Anthony Bowlby, had already argued in 1895 that bacteriology was too specialized to be carried out by unskilled physicians and surgeons.[50]

Pathologists suffered from poor terms and conditions as they had a low status amongst doctors in the late nineteenth century. This is reflected in Andrewes's comment that having had seven years of clinical experience, he had expected to come onto the medical staff at Bart's, having already been on the staff at the Royal Free Hospital. It had never occurred to him to take up pathology as his only walk of life ... [t]o give up all this meant a big change in one's life, but when the offer was made to me I reflected that I loved laboratory work more than Out-Patients, and I finally agreed to take on the job.[51]

Although physicians received fewer than 100 guineas (£105) each per year from Bart's, they could achieve huge sums through private practice. The Bart's practitioners, Sir James Paget and John Abernethy, who worked at the hospital earlier in the nineteenth century, earned in excess of £10,000 a year at their peak.[52] In a meeting with the Medical Council at Bart's in 1895 regarding the proposed salary for the pathologist, the treasurer said: 'Of course £600 a year is not a large or even an adequate payment, but still all that class of work is not highly paid.'[53] Despite this discussion, as already mentioned, Kanthack was only paid £100.

Indeed, in 1902 members of the Medical Council were more concerned with fighting for Andrewes's higher salary rather than arguing for new facilities, although they did compare the existing laboratory unfavourably with those at St Thomas' and the London Hospitals.[54] During that year discussions began regarding new hospital buildings in general, and it was considered that £300,000 to £500,000 needed to be raised for the purpose.[55] In 1903, Andrewes campaigned for the expansion of the laboratories, and in 1904, when matters were discussed regarding temporary accommodation for departments due to the building works, the Medical Council considered it important that new permanent accommodation was found for pathology as the current deficiencies were, once again, affecting student recruitment.[56] When the treasurer and almoners recommended the new block to the governors in 1905, the Medical Council was asked to contribute half of the £20,000 cost.[57] Members of the Medical Council were naturally hesitant about this.[58] Yet, although the endowed hospital was comparatively wealthy, the building of the Pathological Block, the Out-Patients' Block and the growing expenditure of a twentieth-century hospital led to a deficit of £12,000 per year by 1908. Therefore, despite the hospital's apparent wealth, the decision to provide new accommodation for pathology was not a straightforward choice.[59] In 1908, the treasurer (who was by this time Lord Sandhurst) worried about the increasing costs to the hospital, including those resulting from the 'demands of science with all its rapid progress to be constantly met'. He warned that if public contributions could not be increased, half of the beds in the hospital would have to close.[60]

The construction of the Pathological Block finally began in 1907, and it was officially opened in 1909, costing £29,304, including a donation of £1,000 from Mrs Boyd of Essex.[61] The building included spacious laboratories, a library and a post-mortem room.[62] Ten pathologists were employed in the new building – the pathologist, assistant pathologist, a lecturer and seven demonstrators, two of whom were paid for by the medical school.[63] In 1910, the *BMJ* portrayed the pathology department as the most 'complete' in the country.[64]

The physicians and surgeons at Addenbrooke's faced even more difficulties in arguing for funding for pathology than their colleagues at Bart's. As has already been mentioned, Roy was appointed as the first professor of pathology at the University of Cambridge in 1884, but he was an experimental physiologist who undertook some clinical work. From 1885, Roy was awarded £40 a year for his pathology work at the hospital, which was funded by the fees that students paid to Addenbrooke's, which would otherwise have been received by the medical staff. However, Roy's main roles were post-mortems and a few microscopic and chemical tests. An assistant was appointed in 1886 and, after his departure in 1887, the role was formalized with the appointment of a demonstrator. The first person to hold this post was Almroth Wright, who was later famous for his development of the typhoid vaccine and for leading immunological research at St Mary's Hospital.[65] In 1895, the medical staff appointed Ernest Lloyd Jones to the post of Pathologist to the Hospital, although his main duty was to perform autopsies. Unfortunately, the discussion which led to this decision is not recorded within the hospital minutes.[66]

Although Kanthack had been working at Bart's, he had assisted with pathology in Cambridge from 1892, became deputy pathologist in 1895, and finally made a complete move to Cambridge as professor of pathology in 1897, a turn towards the subfield of bacteriology in the university's appointments for this discipline. Kanthack helped Allbutt to set up a small laboratory at Addenbrooke's.[67] In October 1899, Maynard Ward was allocated to those working in pathology, but if the matron thought the space was required for patients, it was to be given up within six hours.[68]

After Kanthack died in 1898, Allbutt and the surgeon George Humphry provided £100 per annum for three years for George Stuart Graham-Smith to carry out pathology work. Only when university teaching interfered too much with Graham-Smith's laboratory work in 1903 did the governors recognize the value of this post but did not see the need to pay for it until 1908. The role was still financed using students' fees.[69] In 1901, Allbutt returned from his summer vacation to find that the room he used for pathology research, testing and teaching had been turned into a matron's room, and £50 worth of apparatus had been thrown into the street. The professor of pathology, German Sims Woodhead, a pioneer of British bacteriology who had established the *Journal of Pathology and Bacteriology*, had given permission.[70] This illustrates the powerful position of the matron at the time rather than Woodhead's lack of regard for laboratory science. In 1908, the staff begged approval for a rearrangement of the Pathological Department. Woodhead was appointed the honorary consulting pathologist, Walter Malden was appointed clinical pathologist and there was another pathologist for tasks such as post-mortems.[71] In 1912, the Addenbrooke's physicians claimed that they had paid £40 per annum for pathology staff over twenty-five

years, and twelve years previously they had built and equipped a laboratory for
£400. In 1914, Malden encouraged a donation of £3,000 and an annual sum for
pathology, which allowed for the building, equipping and maintenance of the
John Bonnett Clinical Laboratory, costing far less than the flamboyant £30,000
laboratory block at Bart's.[72] In the same year, Allbutt reflected on the challenges
he had faced: the 'old Board looked upon their work in the laboratory as an
intrusion of University methods on practical medicine'.[73]

   Clinicians at both Bart's and Addenbrooke's have been shown to have
enthusiastically pressed for funding and specialist services for pathology and
bacteriology despite the unwillingness of lay governors. This was especially the
case for physicians at Addenbrooke's who personally provided almost all of
the early funding. Despite Lawrence's representation of Bart's lay governors as
being interested in character rather than scientific skills, in comparison to the
Addenbrooke's governors they were much more willing to allocate funding for
pathology. In order to examine the practical import of this funding for everyday
clinical practice, the rest of this chapter compares evidence within case notes
from the two hospitals.

Life on the Wards

In 1967, Erwin Ackerknecht called for a 'behaviourist' approach to the history
of medicine – a 'more extensive and more critical analysis of what doctors *did* in
addition to what they *thought* and *wrote*'. He cited an example of research into
case notes which provided an alternative history to that found through textbooks discussing amputations during the Franco-Prussian War (1870–1): a 'lag'
between innovation and application.[74] G. B. Risse and J. H. Warner have more
recently noted the values and limitations of case notes for reconstructing actual
practice in the hospital.[75] This study avoids many of the limitations listed by Risse
and Warner regarding not drawing conclusions from individual case notes by
studying hundreds of case notes, gaining an understanding of their construction.

   A typical case file from Addenbrooke's Hospital has a front sheet summa-
rizing patient details, with the admission examination, medical history and
notes on the patient's stay continuing onto the following page, then another
sheet recording medicines and diet, and finally graphs showing temperature and
weight. Illustrations were commonly used in case notes at both hospitals, for
example pre-printed charts of the body upon which pathological signs could be
marked. However, the case notes for Addenbrooke's Hospital are not uniform
like those created at Bart's. Case notes from Bart's are much better organized
and almost always fully completed. They include a front sheet, including a summary of the case and usually a diagnosis, the notes taken during the patient's stay,
graphs documenting temperature with accompanying charts recording infor-

mation about urinalysis, for example, and a sheet detailing diet and treatments prescribed. Later, extra pages such as pathology and X-ray reports were included. Addenbrooke's case notes sometimes contain similar sources of information, but on many occasions they are not all present.

The case notes at Bart's were not always completed by the physicians who are discussed in this chapter, but often by clinical clerks.[76] Their names always begin with 'Mr' rather than 'Dr' in the internal medicine case notes. However, there are indications that the physicians heavily influenced the composition of the case notes. Geoffrey Bourne's autobiography describes the procedures followed when he was a student. Bourne began his studies at Bart's medical school in 1912 and was subsequently a physician at the hospital. The 'clinical clerks' mentioned in the case notes were medical students in the clinical years of their degree. Clerking involved taking histories and physical examinations but physicians were involved in this process. Bourne describes how John Drysdale led students to describe the condition of the patient through his 'Socratic method of careful questioning', and then led them through 'examination, inspection, palpation, percussion and auscultation'. These notes were read out to the consultants and were criticized by them, as to length, content, style of writing and language.[77]

Despite their creation by students, the physicians' involvement is occasionally illustrated within the case books. For example, the physicians signed requests for the pathologist to undertake Widal's agglutination tests for typhoid (invented in 1896). There are common characteristics in case notes supervised by the same clinician, for example a symbol for typhoid consisting of an oval with a dot in the middle was only used in two case books which documented Dyce Duckworth's patients. The symbol was used often in 1893 (in five out of ten cases) and in 1897 (in two out of four cases).[78] The phrase 'Febris Typhoides' was only used under the care of Tooth for all three cases examined on his ward in 1911 and 1913.[79] This again shows the influence of physicians on their clerks' notes, as it occurred over three years. Some physicians were far more involved in the writing of their patients' case notes. Judging by the handwriting, Samuel Gee appears to have written part or all of sixty-nine out of the ninety sets of case notes attributed to him for pulmonary tuberculosis, diphtheria and typhoid between 1882 and 1899.[80]

The difference between Addenbrooke's case notes and those created at Bart's may be explained by their completion by 'house physicians' rather than 'clinical clerks'. The decision to appoint house physicians, rather than resident students, was made in 1871. From 1872, holders of these posts were paid £65 per annum. Their role was to dispense the medicines which the physicians prescribed, aided by the dispenser, to provide the matron with diet lists for each patient, and to maintain detailed admissions registers including information on the patients' age, disease, parish, any recommendation for admission, and dates of admission and discharge. They also had to give reports on new patients and to present patients who had remained in hospital for over two months at the weekly board meeting. House physicians were also responsible for clinical clerks, the conduct of students and nurses, and post-mortem examinations. The annual reports for Addenbrooke's show that house physicians were usually appointed for one year, but occasionally for up to four years.[81] Therefore it is not likely that the house physicians greatly affected the style of the case notes that they documented over the years, as they were subordinate to the physicians and the turnover was frequent.[82] It appears from the haphazard nature of notes at Addenbrooke's that either standards and expectations for case notes were much higher at Bart's, that students were more conscientious than junior doctors when completing case notes in order to impress their teachers, or even that they were able to take more time to write up their notes.

The sample of case notes from Bart's and Addenbrooke's hospitals is dictated by those which have survived in the hospital archives. A much greater proportion of case notes survived at Addenbrooke's than at Bart's. The vast majority of the Addenbrooke's case notes which are examined in this chapter are for patients who suffered from pulmonary tuberculosis. The examination of these records is particularly significant due to the number of remaining case notes – 1,026 over 40 years. However, many more diphtheria case notes survived at Bart's (234) than at Addenbrooke's (19). The selection of case books which remain at Bart's include occasional case books devoted to the isolation ward where patients with diphtheria were treated, meaning that there are only intermittent years for which diphtheria case notes survive. This is beneficial for this study, however, as the statistics produced are very significant for some years, such as 1896, for which eighty-four case notes from the ninety-two patients with diphtheria have survived (91.3 per cent).[83]

Although there were complaints about patients with infectious diseases being treated at Addenbrooke's Hospital, as mentioned in the introduction, hardly any cases of diphtheria were treated there, and even fewer case notes have survived.[84] The policy regarding admissions at Addenbrooke's Hospital, which attempted to reject diphtheria cases, demonstrates that letters of recommendation from subscribers were not necessary for admission in emergencies.[85] Indeed, Waddington considers that subscriber recommendation became less important from the 1850s, with admission to hospitals based on need.[86] After 1900, when diphtheria cases were admitted for treatment at Addenbrooke's, they were subsequently transferred to a house for convalescence.[87] In 1903, Ernest Lloyd Jones had to write a letter of explanation to the Weekly Meeting committee because he had admitted a patient with diphtheria.[88] Further criticisms of the medical staff's admission of diphtheria cases continued at the weekly meetings. There was concern about incidences of cross-infection.[89] In March 1905 the policy was formalized so that only diphtheria patients requiring immediate tracheotomies were to be admitted.[90] Patients with diphtheria were to be transferred to fever hospitals as soon as possible, whereas at Bart's, patients remained at the hospital.[91]

As the admittance of more infectious patients was not encouraged at Addenbrooke's, this may have affected the general appropriation of bacteriological methods of diagnosis at the hospital. The physicians would be less influenced by the increased status of bacteriology and its worth for diagnosing diseases such as diphtheria and typhoid. However, the pathologists at Addenbrooke's had as much experience with these diseases as those at Bart's through their additional roles in Cambridge. Woodhead undertook bacteriological tests on ice cream for the Borough of Cambridge from 1899, and in 1900 became the consulting pathologist for cases of infectious disease.[92] In 1901, Graham-Smith was appointed as the bacteriologist for the town council, with his position recognized with a salary.[93] Although many bacteriological tests were presumably undertaken by the pathologists at the laboratory at Addenbrooke's, the subject of this chapter is the perceived value of bacteriology and the laboratory for the hospital.

No separately bound pathology forms survive for Bart's or Addenbrooke's, although sheets are often bound in with the case notes. This means that the information gleaned from the case notes is probably an understatement of the number of cases receiving bacteriological diagnoses. However, the entry of bacteriological tests into patient records also demonstrates the significance of the test in diagnosing disease. Another limitation to this study is the variance in the way diagnoses are recorded in three possible places within the case books: in an index, in a space to record 'for what admitted', and as a diagnosis on the sheet listing medications and diet. On many occasions these spaces are not completed, even though the diagnosis is known and is noted right at the beginning of the patient history. Only case notes including a final diagnosis of pulmonary tuberculosis, typhoid or diphtheria, recorded within the case book index or the case file have been examined. Therefore, cases in which bacteriological diagnoses were attempted for these diseases, but the outcome was deemed to be a different disease, are not included.[94] However, an attempt is made to examine the value of laboratory diagnoses by examining the proportion of negative bacteriological tests which still result in diagnoses of typhoid, diphtheria or pulmonary tuberculosis.

Data from case notes needs to be compiled and analysed for a study of over 1,800 case files. Microsoft Access has been an invaluable tool, enabling neat storage of information and cross-comparisons of multiple fields of data. The database was organized by using a 'universal' first page where data such as admission date, physician, disease, information on terms used and other data applicable to all cases were documented. A button from this form linked to three separate forms where data particular to each disease were collected.

Using Bacteriology for Diagnosis

This chapter has so far used administrative records to establish that clinicians were keen to establish specialists in the laboratory. Case notes reveal how physicians used these services. The following extract from *St. Bartholomew's Hospital Journal*, 1896 (see Table 1.1), is the only detailed published record regarding the use of the laboratory at the hospital, and the table was only published for that particular year. It reveals that the majority of bacteriological examinations were carried out in order to diagnose diphtheria, and that during 1895 the number of specimens tested in the pathological laboratory was increasing. Christopher Lawrence has identified Dyce Duckworth and Samuel Gee as epitomizing the ethos of the gentlemen physician and yet they were the most frequent users of the pathological laboratory.[95] However, this extract only shows use of bacteriology in one year, and does not show the proportions of cases of particular diseases which were examined bacteriologically. Also, Table 1.1 is incomparable to the case notes for 1895, as only two diphtheria case notes survive from that year.

Table 1.1 shows that only a small amount of urinalysis took place in the laboratory compared to bacteriological tests. However, the eighty-four case notes in my sample for the following year, 1896, reveal that the urine of sixty patients was analysed for properties such as specific gravity, acidity or alkalinity, and substances such as sugar, phosphates and albumen. Yet Table 1.1 documents only twenty-three urinalyses, indicating either a huge increase in this practice between 1895 and 1896, or more likely that urinalysis could take place more easily outside of the specialized laboratory than bacteriological tests could. Indeed, Joel Howell has discussed the ease of testing specific gravity due to instrumentation, with the urinometer sometimes taken on home visits, and that urinalysis could be undertaken by house officers and interns, and at a later date by nurses.[96] Therefore, urinalysis was almost certainly included in the provision for elementary investigations on the wards, which was discussed during the debate for funding the pathologist.

**Table 1.1: Use of bacteriology at St Bartholomew's Hospital in 1895,
as illustrated in *St. Bartholomew's Hospital Journal.***

PATHOLOGIST'S REPORT

From April 1st to December 31st, 1895, 432 specimens have been sent up to the Pathological Laboratory for investigation. The number of specimens sent up is gradually increasing, as shown by following table.

April	25 specimens	Sept.	41 specimens
May	34 "	Oct.	41 "
June	31 "	Nov.	79 "
July	45 "	Dec.	94 "
Aug.	35 "	No date given	07 "

Specimens were sent up by the following physicians and surgeons:

Sir Dyce Duckworth	55	Mr Willett	**8**
Dr Gee	43	Dr West	6
Dr Champneys	39	Mr Bruce Clark	4
Mr Butlin	32	Mr Lockwood	4
Dr Hensley	28	Mr Walsham	3
Mr Bowlby	25	Mr Vernon	3
Dr Brunton	23	Mr H. Cripps	3
Dr Griffith	18	Dr Moore	1
Dr Church	17	Dr Andrewes	1
Mr Langton	16	Dr Drysdale	1
Mr Jessop	16	No physician or surgeon mentioned	64
Mr Marsh	14		
Mr Smith	8		

These specimens are distributed amongst the various wards and departments as follows:

Radcliffe	99	Coborn	6
Martha	52	Kenton	5
John	28	Rahere	5
Hope	20	Wardmaids	5
Matthew	18	Pitcairn	4
Throat Department	17	Surgery	4
Sitwell	15	Harley	4
Nurses' Home	14	Isolation	4
Elizabeth	12	Lawrence	4
Luke	11	Henry	4
Colston	11	Abernethy	3
Ophthalmic	10	Paget	3.
Darker	10	President	3
Casualty	8	Surgical P.M.	2
Out-patients	8	Faith	2
Stanley	8	Mary	2

Lucas	7	Skin Department	1
Charity	6	No ward mentioned	11
Mark	6		

As to the nature of investigation most frequently required, the following table gives some information:

Diphtheria diagnosis	165	Urine examinations	23
Blood examinations	40	Histological examinations	102
Sputum examinations	19		

Note: Radcliffe Ward was an isolation ward.
Source: 'Pathological Laboratory', *St. Bartholomew's Hospital Journal* (January 1896), pp. 58–9.

A study of case notes thus reveals much more than can be discovered through primary published sources. A total of 405 case notes have survived at Bart's for cases diagnosed with pulmonary tuberculosis, diphtheria and typhoid fever which were admitted between 1880 and 1920, and all of these files have been examined. A larger number of case notes survived from Addenbrooke's from the same period – 1,418 – mostly for pulmonary tuberculosis and typhoid, with only a few examples of diphtheria cases.

Pulmonary Tuberculosis

Tuberculosis was the major single cause of death for British adults in the nineteenth century.[97] It was also a key disease for establishing the science and techniques of bacteriology, as discussed in the introduction.[98] Although the resulting conceptual and practical transition was by no means simple or complete, the identification of the bacillus in 1882 was also striking as it resulted in the disease being reframed as contagious.[99] Records from Bart's for pulmonary tuberculosis have not survived for the period between this date and 1892, meaning that the immediate impact of the bacteriological test cannot be assessed through the case notes. However, as shall be demonstrated in Chapter 2, Gee's private notebooks include extracts from cases at Bart's, which show that bacteriology was used in examining sputum for tubercle bacilli during this time.[100] Culturing of tubercle bacilli was not as immediately popular as the Widal test for typhoid was (see Figure 1.3 on p. 36), but was used in many cases, as Figure 1.1 shows. For the years in which case notes survive for both Bart's and Addenbrooke's, physicians at Bart's can generally be seen to have requested bacteriological diagnosis more frequently than those at Addenbrooke's.

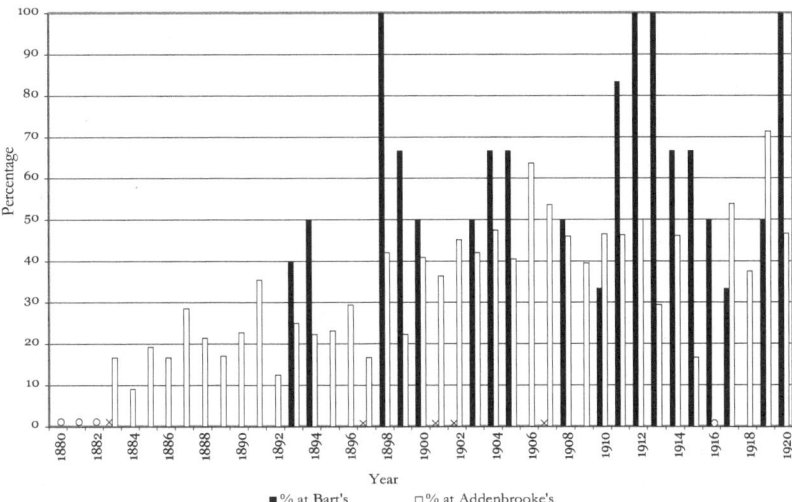

Figure 1.1: Percentage of pulmonary tuberculosis cases in which sputum was examined for tubercle bacilli at St Bartholomew's and Addenbrooke's hospitals. The use of the markers x (Bart's) and o (Addenbrooke's) denote that case notes exist for that year, but do not show any bacteriological tests. These markers distinguish this data from the years in which there are no case notes available for the disease. Source: Medical Casebooks, 1880–1920, Addenbrooke's Hospital Archives, AHPR1/1; Medical Registers, 1881–1920, St Bartholomew's Hospital Archives, MR 16.

Physicians at Bart's and Addenbrooke's did not always accept the results of the bacteriological tests which they ordered. As Stephen Jacyna and Gayle Davis have found in their studies of case notes and pathology reports, clinical diagnoses often took priority. Indeed, false negatives were possible in the laboratory.[101] Table 1.2 shows that in some years a significant proportion of cases only gave negative results in the laboratories at both hospitals and yet diagnoses of pulmonary tuberculosis were still given, particularly at Addenbrooke's in the early twentieth century.

Table 1.2: Cases labelled as pulmonary tuberculosis which received bacteriological tests and those which contained only negative bacteriological results.

Year	St Bartholomew's Hospital		Addenbrooke's Hospital	
	No. of cases tested	Negatives only	No. of cases tested	Negatives only
1880				
1881				
1882				
1883		1		
1884		1		
1885		5		
1886		4		

	St Bartholomew's Hospital		Addenbrooke's Hospital	
Year	No. of cases tested	Negatives only	No. of cases tested	Negatives only
1887			6	1
1888			6	2
1889			7	2
1890			5	2
1891			11	
1892			1	
1893	2		2	1
1894	1		4	1
1895			6	2
1896			5	1
1897			4	
1898	5		8	1
1899	2	1	6	1
1900	1		18	3
1901			20	2
1902			28	8
1903	2		24	1
1904	2		28	7
1905	2	1	15	1
1906			14	
1907			15	6
1908	1		23	2
1909			17	3
1910	1	1	20	1
1911	5	3	19	9
1912	2		15	5
1913	1	1	5	1
1914	2	1	6	1
1915	2	1	1	
1916	2	1		
1917	1		7	1
1918			3	2
1919	1	1	5	3
1920	1		7	2

Source: see Figure 1.1.

However, the physicians did consider that it was important to try to find bacilli in the sputa. For example, at Addenbrooke's in 1889, one patient's sputum samples were tested nine times, with eight negative results. The case file documents that, on 29 November 1889, 'on most careful repeated exams for tubercle bacilli, no typical ones have been found', and then on 26 January 'for about ninth time examd. for tubercle bacilli these found in very large numbers'.[102] At Bart's in 1894, another patient's sputum samples were tested six times with negative results before there was a seventh, positive result.[103] Clearly it was important to try to get that one positive test result.

Wilhelm Roentgen's discovery of the X-ray in 1895 resulted in another diagnostic technology for pulmonary tuberculosis. The potential of X-rays for diagnosis of the disease was suggested shortly after the discovery of the rays. Increasingly, X-rays became the primary diagnostic method for pulmonary tuberculosis, but initially they were used in combination with older forms of diagnosis, because of poor quality images.[104] At the beginning of the 1910s, X-rays were discovered to have distinct advantages over other forms of diagnosis, such as physical signs and tubercle bacilli in sputum, as the early onset of the disease could be spotted in childhood, redefining the aetiology of the disease in the body. It had previously been thought that pulmonary tuberculosis was a disease of adulthood. Within medical journals, C. L. Leonard and R. M. Leslie documented their use of X-rays to reveal that the onset of the tuberculous lesions began deep inside the thorax, and this view was supported by autopsies. X-rays could also show the extent of the tubercle lesions more clearly than any other existing diagnostic method such as percussion of the chest. Despite this, there were still doubts about relying on X-ray imaging in the 1920s, as the evidence could not be supported by any other means during life.[105] The idea of X-rays becoming slowly more popular is an illustration of the concept that new technologies very gradually eclipse each other. Svante Lindqvist has demonstrated that the lives of technologies can be graphically depicted as inverted U-curves, and that these curves overlap.[106] Therefore, from the surviving case notes it appears that the X-ray was first used for the diagnosis of pulmonary tuberculosis at Bart's in 1903, but it was not until 1919 that more X-rays than sputum tests were used in order to diagnose the disease. However, in 1920 no X-rays were recorded for cases of pulmonary tuberculosis. There are only a few surviving records for these years at Bart's, meaning that this information is not conclusive. At Addenbrooke's, despite the Cavendish Laboratory's assistance with early X-rays for the hospital, the case notes do not document their use in diagnosis for pulmonary tuberculosis until 1910. In all but two years, sputum tests were used in a higher proportion of cases than X-rays. In 1915, an equal number of patients had their sputa tested or received X-rays. Only three pulmonary tuberculosis case files remain for 1916, when none of the patients had their sputa tested and only one was given an X-ray.

In summary, Figure 1.1 demonstrates that bacteriology was used more regularly for diagnosis of pulmonary tuberculosis at Bart's than at Addenbrooke's. In addition, X-rays were used at a later date for pulmonary tuberculosis at Addenbrooke's than at Bart's. Therefore it appears that there was a more enthusiastic reception of new diagnostic techniques at Bart's than at Addenbrooke's. For example, even with fifty surviving case notes for 1908, none of the Addenbrooke's case notes record an X-ray screening for pulmonary tuberculosis, even though Field had learned how to use the X-ray machine in 1907.

Diphtheria

By far the largest number of tests documented in the Bart's Pathological Laboratory report for 1896 (Table 1.1) were for diphtheria. One of the most characteristic symptoms of diphtheria is the formation of a false membrane in the throat which could choke the patient, and the word 'diphtheria' is derived from the Greek for membrane.[107] However, the disease affects the blood as well as the throat. Death can result from either asphyxiation, or from toxaemia, meaning that the performance of a tracheotomy does not always equal recovery.[108] There were additional complications, as in a 'significant minority of patients' the membrane does not form, meaning that there were problems in diagnosing the disease before bacteriology.[109] This is demonstrated by patients treated at Bart's who were admitted after the time at which bacteriological diagnosis started to take place at the hospital. In fifteen of the extant case files for patients admitted after 1896, the bacteriological test was positive, yet there was no membrane present. Diagnosis could also be difficult as the membrane can be a symptom of croup, scarlet fever or measles; it was difficult to differentiate diphtheria from other diseases affecting the throat such as thrush and quinsy; and because diphtheria has many symptoms which are not always present.[110] Therefore, bacteriology should have been very useful for diagnosis of the disease at Bart's and Addenbrooke's. Indeed, historians of diphtheria in Australia have discovered that the bacteriological diagnosis of diphtheria became so 'determinative' that children who did not receive positive diagnoses were sent home, some cases ending fatally.[111]

Although the diphtheria bacillus was isolated in 1883, the reliability of this claim was questioned as so many false announcements of a discovery of the bacillus had been made. Loeffler's discovery did not result in a consensus from his peers, and the search for a bacillus still continued. Between 1886 and 1890, his research was confirmed by others.[112] There was also reason to doubt bacteriological tests, as early preparations of diphtheria cultures were not stained, and so were 'open to grave chances of error'.[113] Figure 1.2 shows that diphtheria patients did not receive bacteriological diagnosis until over ten years after the isolation of the bacteria. For Bart's, ninety-seven records survive for 1890, and thirty-eight survive for 1891, with no bacteriological diagnoses being recorded. This is not surprising considering the difficulties with the reception of the identification of the bacillus and the problems with unreliable test results. Only two diphtheria case files remain for 1895, with both including bacteriological tests, but the paucity of data means that this is not conclusive. However, bacteriology had certainly become routinely used for diagnosis of diphtheria by 1896, when the test was utilized for seventy-two out of eighty-four cases. The hospital's 'Statistical Report' shows that there were ninety-two cases of diphtheria admitted in 1896, so this is a very significant sample.[114] When the test began to be used, it was recorded in the case notes more frequently than the clinical diagnostic test of the presence of the false membrane in 1895 and 1896.[115] It seems that the Bart's physicians wanted to be sure about the validity of this test before using it wholeheartedly.

The test may have been popular from the mid-1890s as bacteriological tests for diphtheria did not produce negative results as commonly as those for tuberculosis. Just as for tuberculosis, there was often a mixture of positive and negative test results, but this was frequently because patients were tested to discover if they were still infected with the bacilli before they were released. For example, one case at Addenbrooke's received seven positive diagnoses from ten tests.[116] However, in general the diphtheria case notes were completed very poorly at Addenbrooke's, with the majority being discharged straight away to the separate fever hospital, and so they cannot be very usefully compared with the Bart's data. It is unfortunate that there are no records for diphtheria from between 1892 and 1894 for either Bart's or Addenbrooke's, meaning it cannot be seen how quickly the test was adopted after Loeffler's discovery had been confirmed.

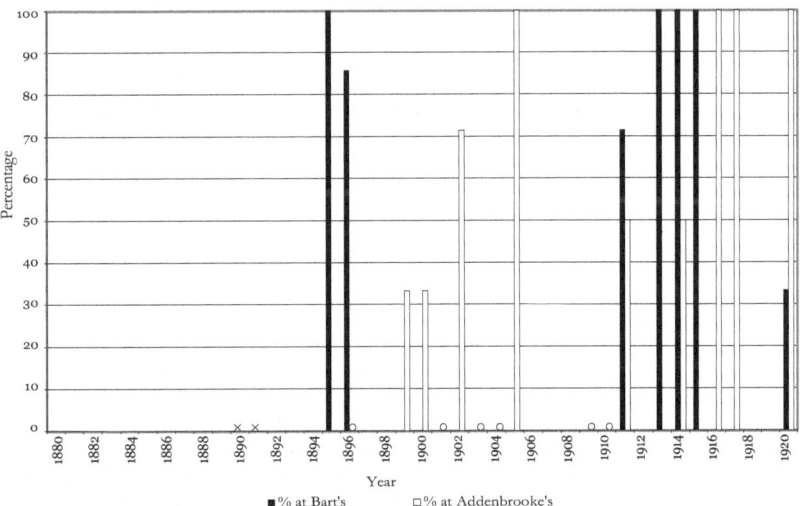

Figure 1.2: Percentage of diphtheria cases which received bacteriological testing at St Bartholomew's and Addenbrooke's hospitals. Source: see Figure 1.1.

The sudden adoption of the test can also be related to the successful therapy which was created in the late nineteenth century. The diphtheria bacillus releases a poison into the blood, which could be obtained in the laboratory by culturing bacteria. In 1890, Emil von Behring injected this poison into healthy animals. The animals produced antitoxin in the blood serum as an immune response to the effects of the toxin. Fairly unsuccessful clinical trials ensued, delaying the widespread use of the therapy, before more impressive effects were gained after an increase in the dose.[117] Although an immunological therapy, antitoxin could not have been created without the isolation of the bacteria which is needed in its production. At the meeting to discuss the appointment of the pathologist at Bart's in January 1895, Shore mentioned antitoxin as the reason why the public now recognized the importance of bacteriology.[118] Antitoxin was used in 95.24 per cent of surviving Bart's records for diphtheria in 1896, showing an almost universal acceptance of the worth of this drug, soon after its introduction into medical practice. As mentioned, there were hardly any cases of diphtheria treated at Addenbrooke's, with the maximum in any one year (1902) being seven. However, from 1896 (the first year for which records of diphtheria survive), at least half of the patients at Addenbrooke's received antitoxin, except for the years 1916, 1921 and 1923, when only one case per year was admitted.

The case notes from Bart's support Anne Hardy's conclusions regarding diphtheria. Utilizing published sources, Hardy has observed a lack of change from the treatments of tracheotomy to intubation in England, in contrast to the almost complete transition to the latter in the USA by 1891.[119] During the period 1890 to 1920, the case notes record that only 3 intubations took place at Bart's, compared to 108 tracheotomies from 234 surviving records. Similarly, within hospitals run by London's Metropolitan Asylums Board (MAB), tracheotomies were used more commonly than intubation into the 1920s.[120] Practices changed in other areas of Western Europe as antitoxin stifled the growth of the membrane, taking away the need for tracheotomies. The less drastic procedure of intubation was used in conjunction with antitoxin.[121] However, physicians at Bart's and the MAB either did not recognize the safer treatment which could be used with antitoxin, or the patients were so unwell when they came to the hospital that tracheotomy was necessary. Many cases in the Bart's records follow this pattern, for example the folllowing case from 1896, where a tracheotomy was performed on the day of admission:

> On adm ... breathing rapidly and with much difficulty ... stridor – considerable cyanosis ... Temp = 103° ... Soon after adm. patient was given an injection of Liq. Strych miii, a hot bath and a poultice applied to throat. The dyspnoea and stridor became rapidly worse, and at 8.15pm tracheotomy was performed.[122]

Other measures were taken, but eventually a tracheotomy seemed to be necessary for this patient, even though he was under the care of Thomas Lauder Brunton, an enthusiast of laboratory sciences and new technologies.[123]

To summarize, although there was a slow response to scientific discovery in terms of diagnostic practice, the utility of the therapy of antitoxin was recognized very quickly. Bacteriological diagnosis at Bart's may also have been quickly adopted in the mid-1890s in order to decide whether a patient required continued treatment with the serum and to see if the patient had been cured and could be safely released from the hospital. However, the study of tracheotomies versus intubation indicates that Bart's staff members were not always quick to adopt different techniques, although the emergency nature of the admissions may explain the continued use of tracheotomies.

Typhoid

Bacteriology had a significant impact on the diagnosis of typhoid fever as the disease is difficult to diagnose clinically; early symptoms are similar to those of malaria, hepatitis, tuberculosis, brucellosis and typhus, amongst other diseases.[124] In the late nineteenth century two ways were devised for diagnosing typhoid fever using experimental methods. The first, the Ehrlich test, was invented in 1882, and the second, the Widal test, in 1896. The first test, which was called the 'Diazo reaction' by Paul Ehrlich, involved mixing urine with chemicals, which produced a carmine colour if the patient had typhoid. The problem was that this test also produced a positive reaction for other diseases, such as malaria and tuberculosis. The Widal test involved waiting for blood or serum to agglutinate in masses when mixed with the typhoid bacillus. This only happened if the patient had typhoid (but as is discussed in Chapters 5 and 6, could be complicated if a patient had been inoculated for the disease). However, this test does not work in the first week of the illness.[125]

Joel Howell has examined the use of the Widal test in two American hospitals from 1900 to 1925. He found that 93 per cent of patients received the test at the Pennsylvania Hospital in 1900, followed by 93 per cent in 1909, 50 per cent in 1920 and 100 per cent in 1925. In 1900, 100 per cent of patients at the New York Hospital received the test, and 85 per cent received it in 1910.[126] The test appears to have been at least as well received at Bart's, where in the first year after records survive following the development of the Widal test, 100 per cent of patients received the diagnostic test (as shown in Figure 1.3). This shows that Bart's physicians were not at all conservative in adopting this new technique in comparison to these American hospitals. It was not until 1903 that the Widal test was used as routinely at Addenbrooke's, even though the test was first used there in 1897, compared to 1898 at Bart's.

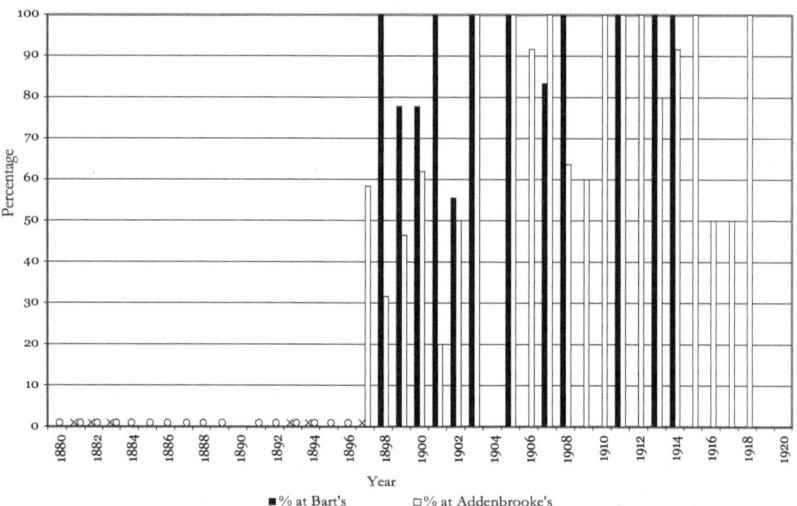

Figure 1.3: Percentage of typhoid cases which received the Widal test at St Bartholomew's and Addenbrooke's hospitals. Source: see Figure 1.1.

Supporting Lindqvist's theory that older technologies are gradually eclipsed, Ehrlich's reactions were still carried out after the development of the Widal test. The first Ehrlich tests in the extant records from Bart's (there are no typhoid records which survive from 1884 to 1892) were performed in 1893, when seven out of ten of Duckworth's cases received the test. After 1896, when the Widal test was invented, the Ehrlich test was used once out of four cases in 1897 by Duckworth and once out of eight cases in 1898 by Church, so there was a transition period before the use of the Widal test became the only one of these two methods to be used. However, at Addenbrooke's, the use of the Ehrlich test continued until at least 1916, and Figure 1.4 shows that it was performed at least as much as the Widal test until 1907.

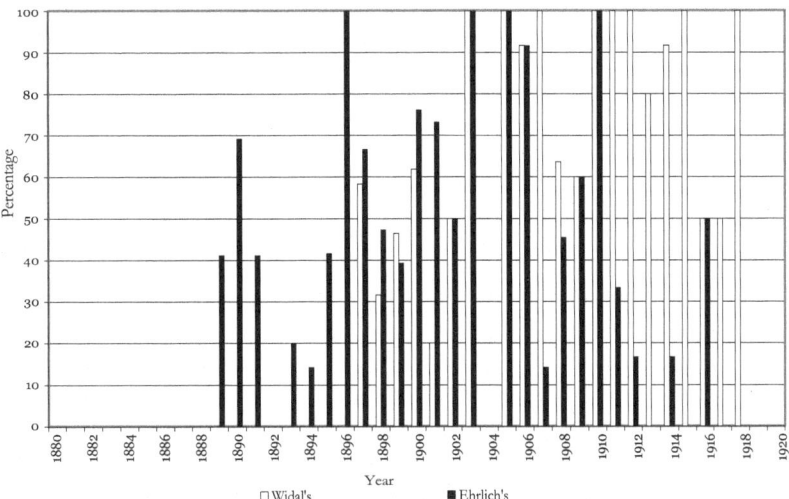

Figure 1.4: Percentage of typhoid cases which received Ehrlich and/or Widal tests at Addenbrooke's Hospital. For all years, apart from 1904, 1919 and 1920, typhoid case notes survive. Source: see Figure 1.1.

There were reasons for the continued use of the older method. Although the Ehrlich test was nowhere near as accurate as the Widal, it was a test which could be carried out without recourse to the pathology laboratory, being listed within the case notes amongst the results of urinalysis. However, the Widal test was also problematic as it was not useful in the first week of onset of typhoid.

The number of cases which only produced negative Widal test results, but were still labelled as typhoid, was not as high in proportion as those for pulmonary tuberculosis. However, at Addenbrooke's in 1901 and 1902, at least half only tested negatively, and in 1915 and 1917 when there was only one case each year, both of these cases only received negative test results. Yet, generally it seems to have been a much easier process to attain positive results from a Widal test than for tubercle bacilli in sputa, with only a maximum of three negatives before a positive test at Bart's, and only two negatives before a positive test at Addenbrooke's, perhaps explaining why this was a much more popular diagnostic tool than culturing tubercle bacilli.[127] Bacteriology may also have been utilized more for typhoid than for pulmonary tuberculosis because of the relative difficulty of diagnosing typhoid before the tests were introduced, in comparison with the pre-existing methods of percussion and auscultation, along with common symptoms such as haemoptysis and wasting, which were used for the diagnosis of tuberculosis. Additionally, at Bart's there was more expertise available to carry out these tests, as the use of these diagnostic technologies coincides with Kanthack's appointment as pathologist in 1895. In contrast, at Addenbrooke's the governors were reluctant to fully fund a specific hospital pathologist with bacteriological interests, with the post shared with the university, perhaps accounting for the delay in the routine use of the tests.

Comparisons and Conclusions

Figures 1.1 to 1.3 demonstrate that laboratory methods of diagnosis, conducted by a specialist in the laboratory, were in general more routinely used at Bart's than at Addenbrooke's. The key difference in policy between the two hospitals appears to have been the retention and treatment of diphtheria patients at Bart's. Hence, the high incidence of diphtheria patients resulted in more persuasive arguments for the establishment of the laboratory. The increased status which antitoxin brought to laboratory medicine from the early to mid-1890s provided promise for diagnosis and treatment of other diseases, and could have influenced the faster adoption of the Widal test at Bart's in comparison to Addenbrooke's, with the test developed shortly after the discovery of diphtheria antitoxin.

Another factor which could have affected the use of bacteriological testing is the decreasing importance of subscriber recommendation from the 1850s, so that admission to hospitals was based on need.[128] Tuberculosis patients, as non-emergencies, were more likely to have received medical attention and bacteriological tests before admission, perhaps even as justification for hospitalization. Emergencies entering the hospital, including acute cases of typhoid or diphtheria, where payment by poor law unions or support from a subscriber was not considered such an important issue, were much less likely to have seen a doctor before admission to the hospital. Therefore in addition to more accurate clinical methods of diagnosis for tuberculosis, this may also explain why less bacteriological tests took place.

The results of the data collection may be skewed by the representation of individual physicians in the case books throughout the period studied at each hospital. As shown in Table 1.3, some physicians are represented by a much larger percentage of the surviving case notes than others at both hospitals. However, an examination of a sample of these doctors shows that none particularly distort the trends for each hospital. Generational differences do not appear to have played a significant role in the decision to adopt new methods. Older physicians at Bart's, such as Church (b. 1837) and Gee (b. 1839), both undertook tests in a large percentage of cases, further demonstrated by the early acceptance of bacteriological testing at the hospital. At Addenbrooke's, there is no noticeable difference throughout the period between the acceptance of bacteriology by older or younger physicians, demonstrating that the tests were accepted institutionally rather than individually. However, there is a slight difference between the acceptance of testing sputum for tubercle bacilli in the late nineteenth century by Peter Wallwork Latham, born in 1832, in comparison with Bradbury, who was born in 1841, and was a younger physician with a strong interest in experimental science. For Bart's, where fewer case notes survive, it could be an issue that certain physicians' notes are the only ones remaining for many of the years studied. However, this is only the case in thirteen out of forty years, and different doctors are represented within the case notes from these years.[129]

Table 1.3: Number of case notes per physician for pulmonary tuberculosis, diphtheria and typhoid at St Bartholomew's Hospital and Addenbrooke's Hospital, 1880–1920.

St Bartholomew's Hospital

Physician	Number of cases
Samuel Gee	90
Dyce Duckworth	58
William Church	51
James Andrew	45
Thomas Lauder Brunton	42
Howard Henry Tooth	33
Philip Hensley	31
Norman Moore	14
Percival Horton-Smith Hartley	7
James Calvert	6
Wilmot Herringham	5
Samuel West	4
John Drysdale	3
Walter Langdon-Brown	1
James Matthews Duncan	1

Addenbrooke's Hospital

Physician	Number of cases
John Buckley Bradbury	426
Lawrence Humphry	409
Peter Wallwork Latham	170
Donald MacAlister	154
Ernest Lloyd Jones	142
John Aldren Wright	100
T. Clifford Allbutt	23
George Haynes	2
Arthur Cooke	2

Source: see Figure 1.1.

This chapter has presented an uncomplicated story of how bacteriology and the laboratory were received enthusiastically at Bart's. Bacteriological methods of diagnosis were used more regularly at Bart's than at Addenbrooke's and the Bart's laboratory received far more funding at an earlier date. However, the smaller number of case notes from Bart's and the more thorough completion of the case notes for that hospital means that this is a tentative conclusion. Yet the governors at Bart's were certainly more helpful than those at Addenbrooke's in establishing the laboratory and its staff, enabling more emphasis on bacteriological diagnosis. Therefore a comparison of this study with recent representations of Bart's appears to reveal that the reception of specialist diagnostic procedures by elite physicians has been misunderstood in the historiography. Chapter 2 will show that these representations have been influenced by accounts from Samuel Gee's students, and that these have been repeated by doctors and historians in their histories of the hospital. Yet this revision of the historiography is complicated as there are also indications that physicians at Bart's were torn between conservatism and adoption of new technologies and specialization, and this may account for some of the representations put forward in the last fifty years.

2 INTEGRATING THE LABORATORY INTO GENTLEMANLY PRACTICE

> Gee's opinions on one subject, books or men, rarely last the same for any time. One would have talk with him, and a fortnight after resume the conversation at the point where it was left off and then it would be found that the whole tone of his thought had changed ... We cease to change only when we cease to exist.
>
> J. Wickham Legg (1911)[1]

In order to understand the integration of the laboratory into practice at an elite London hospital, this chapter analyses the language used within the case notes and presents two biographical studies of physicians at Bart's. The purpose is to understand why some published representations of physicians' attitudes towards the laboratory and its specialists differ from their practice between 1880 and 1920. Although the chapter mainly focuses on Bart's, comparisons are made with the Addenbrooke's case notes, utilizing the same dataset as Chapter 1.[2] First, the transition in the use of the word 'natural' to the word 'normal' within the case notes is analysed. This shift illustrates how patterns in language associated with laboratory science and quantification complicate the story of physicians' reception of bacteriology. Second, the chapter re-examines and revises the representations and self-representations of physicians from two generations at Bart's who have already been discussed in detail by historians. The first is Samuel Gee, physician at Bart's from 1868 to around the turn of the century.[3] The second is Thomas Horder, physician at Bart's from 1912 to 1936.[4] This section focuses on their individual use of bacteriology as revealed within the hospital case notes, and Gee's unpublished notebooks. Their publications are also discussed and placed within the context of their views on specialization and laboratory science in general. Finally, the chapter questions how the public perceived specialization and bacteriology at this time.

Natural or Normal?

The word 'natural' was not simply a precursor of the term 'normal'. Waltraud Ernst argues that the transition between the terms indicated a swing 'from a religiously ordained natural order to a scientifically grounded secular framework ...

an important shift in kind and semantics and not merely one of magnitude and terminology'.[5] Therefore the use of the terms in order to describe the body may be a valuable indicator of the mentality of physicians towards laboratory science. 'Natural' referred to disease as non-specific, and as 'systemic imbalances in the body's natural harmony'. It was related to an individual's well-being, which could differ with the seasons, or with ethnicity, location or sex, for example.[6] By the twentieth century, the term remained in parlance, with its use influenced by its derivation from 'nature' in order to consider 'moral behaviour and standards' such as sexuality.[7]

Until the middle of the nineteenth century, the word 'normal' was defined as 'standing at right angles', whereas by the twentieth century it had social and moral implications: a standard against which to be measured.[8] The term has been argued by Ian Hacking to be the 'most powerful ideological tool of the twentieth century' and when linked with patient behaviour, particularly with the derivation, 'normative', its use within medicine has been critiqued by philosophers and sociologists.[9] How did such a transition occur? Norms were defined in various areas of science, mathematics and medicine. 'Normal' was used by experimental physiologists in the second half of the nineteenth century, when 'universalized norms defined by laboratory science' became the way to consider well-being. These norms could be depicted in the graphical and quantitative representation of temperature, pulse, respiration and urinalysis – matching measurements to 'normal standards'.[10] Machines used in experimental physiology such as the kymograph, recorded graphical images of the pulse or the respiration of an animal, and the recording of quantitative information in a graphical form subsequently became prevalent in clinical medicine.[11] In 1865, the leading French physiologist Claude Bernard argued in his manifesto that 'knowledge of pathological or abnormal conditions cannot be gained without previous knowledge of normal states'. He referred to these 'normal states' as being 'laws' or amounting to 'determinism'.[12] In the 1960s, Georges Canguilhem expertly situated Bernard's work in its nineteenth-century context, but also critiqued Bernard's positivist, quantitative definitions from a 1960s point of view, insisting that changes in the body should be described qualitatively.[13]

The influence of the idea of 'normal' had wider medical connotations with earlier influences.[14] In the 1830s, the Belgian statistician Adolphe Quetelet collated measurements and other numerical data in order to judge the average physical type: 'the average man'.[15] Norms in medicine can be seen to have an even longer history within the huge hospitals in Paris in the late eighteenth and early nineteenth centuries, which allowed the collation of large amounts of data on diseases. Patients were treated as an 'example of the universal' within these huge wards, examined in life and by autopsy, providing comparative data which resulted in increased production of knowledge regarding disease.[16]

One of the major findings of John Harley Warner's pioneering study of patient records from the Massachusetts General Hospital and the Commercial Hospital of Cincinnati is the change of language from 'natural' to 'normal'. He shows, with a study of about 4,000 case notes from the two American hospitals, that there was a significant transition from the use of the word 'natural' to describe a state of well-being to the term 'normal' by the mid-1870s.[17] Noticeable differences can be seen between Bart's in London and the two hospitals in the USA. Hardly any of the American records included 'natural' by the 1870s and 1880s.[18] Yet, Figure 2.1 reveals that 'natural' was in use at Bart's throughout the period 1880–1920.

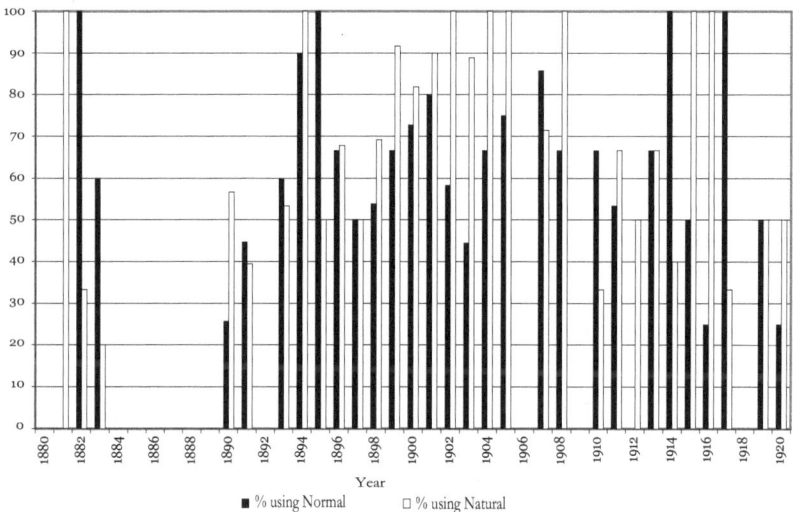

Figure 2.1: Percentage of case notes in which the terms 'natural' and 'normal' appear at St Bartholomew's Hospital. Source: Medical Registers, 1881–1920, St Bartholomew's Hospital Archives, MR 16.

As 'natural' and 'normal' were used concurrently at Bart's, it is necessary to examine the case notes further to examine which word was used more frequently; this is because there is the possibility that the terms were only used very occasionally within each case note. Each of the 405 individual Bart's case notes was examined, revealing which word, 'natural' or 'normal', was used the most within each case. Figure 2.2 represents this analysis, illustrating that although the predominance of the term 'natural' fluctuated, in general it was used more than 'normal' for much of the first two decades of the twentieth century.

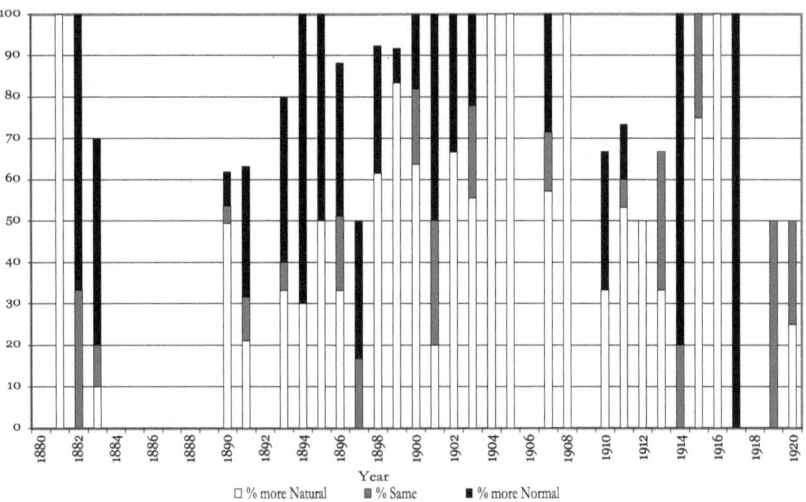

Year

□ % more Natural   ■ % Same   ■ % more Normal

Figure 2.2: Percentage of case notes in which one of the terms 'natural' or 'normal' is used more than the other at St Bartholomew's Hospital. Does not always equal 100 per cent as not all cases include the words 'natural' and 'normal'. Source: see Figure 2.1.

The word 'natural' was evidently very much in use at Bart's throughout the period 1880–1920. Warner has argued that the use of language in medical records reflects the change in 'cognition' of the physicians.[19] Language at Bart's is both significantly different to the American hospitals and to that used in Cambridge. At Addenbrooke's, as would be expected at a hospital with associations with the famous physiology department at the University of Cambridge, 'normal' superseded 'natural' from an early stage, as illustrated by Figures 2.3 and 2.4. Figure 2.3 simply reflects whether the terms were used in each of the case notes and, as with Figure 2.2, Figure 2.4 presents the results of a more detailed analysis of which term was used more frequently within each of the 1,418 case notes.

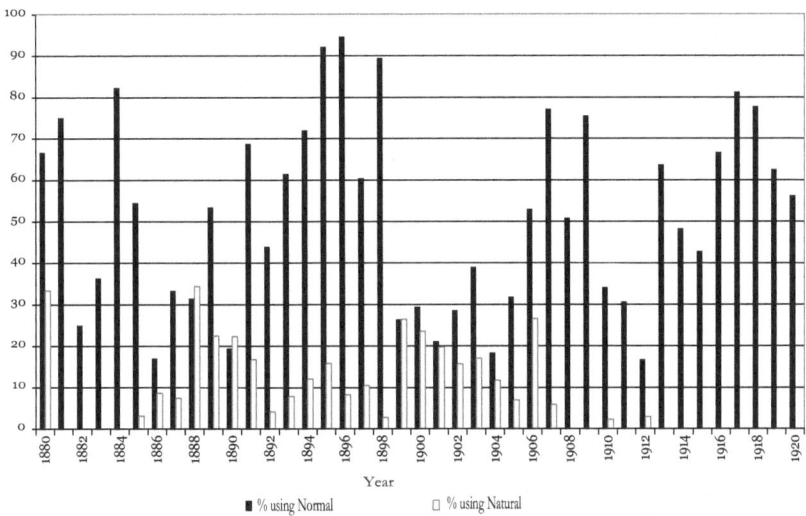

Figure 2.3: Percentage of case notes in which the terms 'natural' and 'normal' appear at Addenbrooke's Hospital. Source: Medical Casebooks, 1880–1920, Addenbrooke's Hospital Archives, AHPR1/1.

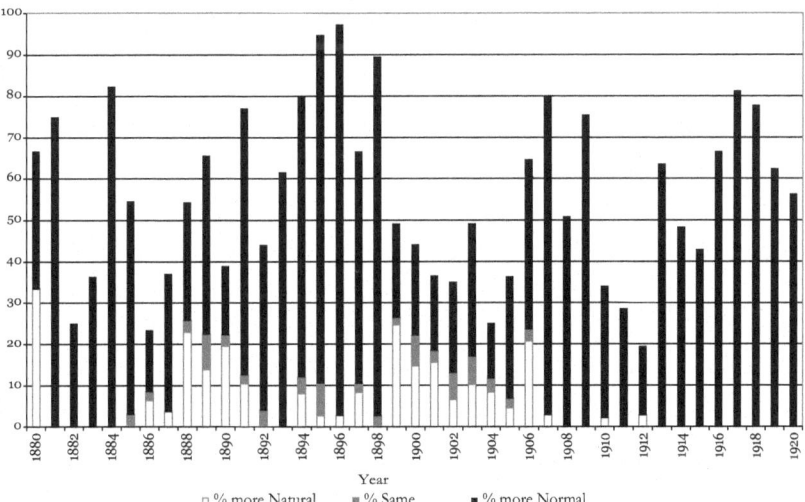

Figure 2.4: Percentage of case notes in which one of the terms 'natural' or 'normal' is used more than the other at Addenbrooke's Hospital. Does not always equal 100 per cent as not all cases include the words 'natural' and 'normal'. Source: see Figure 2.3.

Many different physicians compiled the case notes which have been analysed, as illustrated in Table 1.3. Therefore it is worthwhile examining a selection of physicians in order to determine whether individuals affected particular use of language. A sample of six Bart's physicians born between 1837 and 1876 demonstrates that physicians of various generations all continued to use the term 'natural' throughout the period studied. Indeed, case notes supervised by the younger physicians, Howard Tooth and Percival Horton-Smith Hartley, used the term 'natural' more often than 'normal'. Lawrence explores how the laboratory could be fitted into a patrician world through a study of Horton-Smith Hartley. This physician conducted experimental research and published his experimental work on typhoid. However, he also wrote about morbid anatomy, a subject valued by gentlemen physicians, and had wide interests including publishing a book on medieval history. Lawrence believes that Horton-Smith Hartley's experimental work was compatible with the gentlemanly tradition as it was a solo project.[20] The language shown within Horton-Smith Hartley's case notes appears to reveal a younger physician who is heavily influenced by old-fashioned terminology. In contrast, Samuel Gee, a much older physician, used 'natural' and 'normal' in fairly equal proportions at a much earlier time. As discussed in Chapter 1, for some years only one physician's case notes are represented by the remaining case notes. Analysis of individual physicians' notes demonstrates that patterns appear to be institutional rather than individual. No single physician appears to have skewed the results for Bart's, with the term natural used by all throughout the period studied.

Addenbrooke's physicians all used 'normal' frequently throughout 1880–1920, following the same pattern shown in Figures 2.3 and 2.4. This is even the case for the physicians who were born in the 1830s, although this generation includes Allbutt, a strong supporter of continental experimental science. Therefore, patterns are generally followed by all at each of the hospitals, regardless of the physician's age and so show mentalities of a group working in a particular environment, rather than those of individuals educated at different times in the development of biological and mathematical sciences.

In order to qualitatively understand the use of terminology in the case notes, it is valuable to analyse the specific functions of the body which the words are used to describe. 'Natural' and 'normal' were often written close together in single case notes, even for the same types of descriptions. For example, in an Addenbrooke's case file from 1902, a sentence was written: 'Cor. [heart] natural and in normal position'.[21] A case note from Bart's stated in 1896, 'Knee jerks nat.' and then in the same handwriting on the next day, with only two lines in between, was written, 'Knee jerks normal'.[22] This apparently random use of the terms could be employed to question the significance of the change in use of 'natural' to 'normal', especially for Bart's where both words were commonly utilized until 1920. However, these examples relate to states which could perhaps be subjective rather than objective. Most of the other instances where the terms are both used within the same case file include a reference to the temperature being normal.

Purely quantitative measurements quickly became referred to as 'normal'. Indeed, Warner found with the American case notes that quantitative measurements were described as 'normal' before qualitative descriptions. This was especially the case for evidence gained using instruments such as the 'stethoscope, thermometer, or time-piece'.[23] Figure 2.5 shows the percentage of instances in which temperature is referred to as either 'natural' or 'normal' at Bart's. It can clearly be seen that temperature was nearly always referred to as 'normal', right from the beginning of the study. At Addenbrooke's, temperature was never referred to as 'natural' in the sample case notes. Bart's follows a different pattern when qualitative heart sounds are examined (Figure 2.6). Temperature was easier to consider as objective and 'normal' than heart sounds; within the case notes, a printed line drawn across the graph paper for the recording of temperature designated normal body temperature. In contrast, Addenbrooke's physicians predominantly used 'normal' to describe heart sounds, right from the beginning of the period, and exclusively so from 1903.

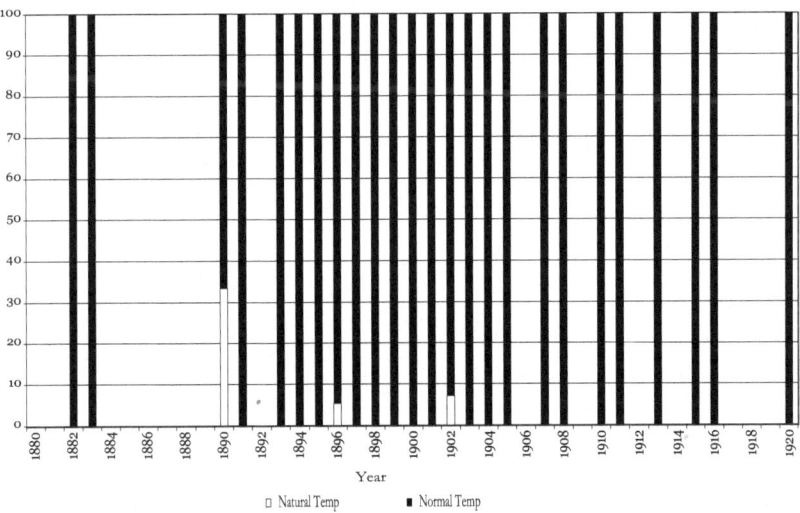

Figure 2.5: Percentage of instances in which temperature is described as 'natural' or 'normal' at St Bartholomew's Hospital. Source: see Figure 2.1.

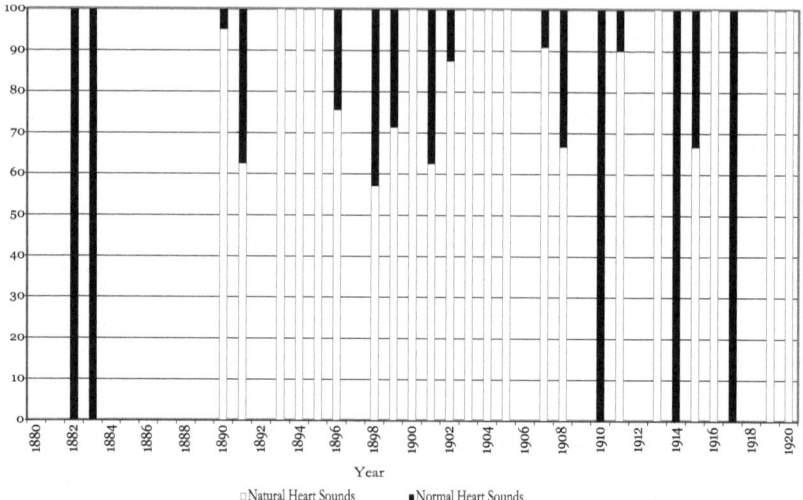

Figure 2.6: Percentage of instances in which heart sounds are described as 'natural' or 'normal' at St Bartholomew's Hospital. Source: see Figure 2.1.

Bart's physicians and students also used natural to describe organs and parts of the body: the shape, position and what was palpable (Figure 2.7). Yet again, Addenbrooke's closely followed the American model, even for this kind of subjective description (Figure 2.8). This analysis of the way in which 'natural' and 'normal' were used in general and in particular contexts has revealed a stark disparity between the two hospitals. Therefore, further explanation of how laboratory science was perceived at the two hospitals will assist in understanding these differences.

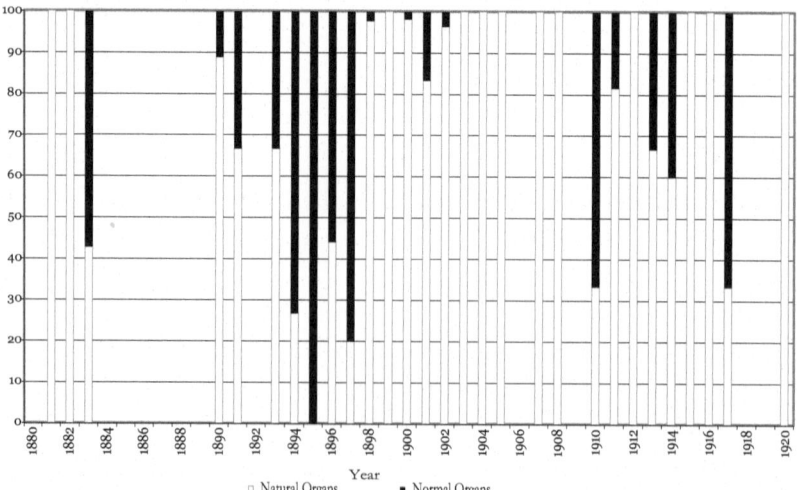

Figure 2.7: Percentage of instances in which qualitative descriptions of organs or parts of the body are described as 'natural' or 'normal' at St. Bartholomew's Hospital. Source: see Figure 2.1.

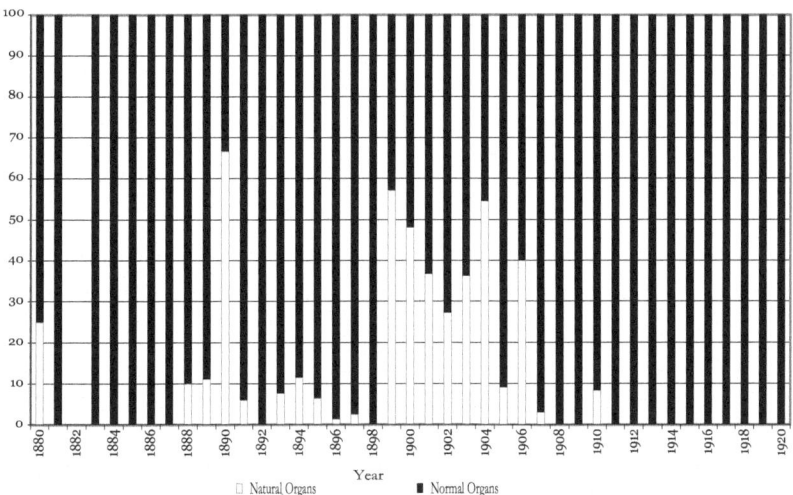

Figure 2.8: Percentage of instances in which qualitative descriptions of organs or parts of the body are described as 'natural' or 'normal' at Addenbrooke's Hospital. Source: see Figure 2.3.

By the late nineteenth century, the University of Cambridge was arguably the 'leading centre of physiological research in England, if not the world'.[24] The influence of experimental physiology may point to why changes in language were so different at Addenbrooke's in comparison to Bart's. Warner frames the change from 'natural' to 'normal' as coinciding with the gradual growth of experimental physiology as a model science for medicine. The intended aim of therapeutics changed from acting on the body as a whole in order to restore the patient's individual 'natural' balance, to working on specific parts of the body in line with knowledge gained from experimental physiology. The goal became to return the body to universalized 'measured, quantified norms' determined or influenced by the laboratory.[25] Worboys has described how bacteriology superseded physiology from the 1870s, with medical understanding having developed from concentration on the results of disease in the earlier nineteenth century, to disease processes in the mid-nineteenth century, to causes at the end.[26] Therefore, it is important to understand how the predominant medical science for understanding theories of disease causality prior to bacteriology influenced physicians' thoughts about disease and diagnosis. This would have affected how receptive physicians were to more theories which were derived in the experimental laboratory. This is particularly relevant for a comparison with the hospital which was associated with the University of Cambridge. Additionally it is hard to separate the influence of experimental physiology and bacteriology on concepts of norms in the late nineteenth century as even the forerunners of bacteriology – concepts of 'specific chemical, fungal, and animalcular causes of disease' – also had an effect on ideas of universalized and specific disease causation from the 1840s.[27]

Mid- to late-nineteenth-century German experimental physiology is associated with 'biophysics' and the use of instruments (and also chemistry), sometimes resulting in graphical representation such as the recording of the pulse by the kymograph and sphygmograph.[28] This influence can clearly be seen in the case notes of the two English hospitals, where there is almost always a graph in each case file showing temperature. Chemical assaying, of urine for example, also showed the influence of chemical physiology. The British style of experimental physiology was not as exacting as in Germany, with less emphasis on graphical representation, and more on the senses. In 1874, however, Foster did receive money for recording apparatus for his laboratory which may have been a kymograph. Walter Holbrook Gaskell, one of Foster's protégés, spent a year working with Carol Ludwig in Leipzig in 1874. Ludwig chiefly created the kymograph. Other students from Cambridge also spent time in German laboratories. In Leipzig, Gaskell had experimented on a dog with the kymograph in order to measure blood flow, whereas in Cambridge he used a frog, 'as if in imitation of his smaller and less elaborate surroundings'. By 1883 a kymograph was definitely available in Cambridge, as Gaskell used it for an experiment on the heart of a tortoise.[29]

The importance of physiology and the use of experimental and quantitative methods may have had a direct impact on the differing use of vocabulary at Addenbrooke's Hospital compared to Bart's. Huge numbers of students studied elementary physiology in Cambridge, with an intake of 130 in 1885, and 190 in 1889. Foster, Humphry and Allbutt thought it was preferable for students to have training in pre-clinical natural sciences, rather than the old system where medical students went straight from the mathematical tripos to clinical studies.[30]

Practical physiology teaching developed in England in the late 1860s, beginning with Foster's elective course at University College London (UCL). Foster remembered his mentor, William Sharpey at UCL, having to use a cylinder hat to explain the kymograph in the 1850s in the absence of this equipment.[31] The importance of physiology for the medical curriculum was acknowledged in 1870 when the Royal College of Surgeons introduced a requirement that thirty classes of 'practical' anatomy and physiology should be a requirement of the licentiate, with students themselves partaking in experiments. The University of London adopted this requirement in 1871. Geison argues that this development led to the growth of physiology in Britain. The discipline expanded between 1870 and 1900, arresting the flow of British physiologists to the continent.[32]

Although physiology at Bart's did not have the reputation of the Cambridge school of physiology, Thomas Lauder Brunton was very enthusiastic about experimental pharmacology and physiology, publishing widely and working on treatments which had practical applications such as the action of digitalis. Like contemporaries at Cambridge, he spent time in laboratories at UCL and in Vienna, Berlin and Leipzig in the early 1870s. However, rather than specializing in experimental physiology like Gaskell, he combined research with hospital medicine.[33] Yet physiology at Bart's was reluctantly taught, although many of the initial problems physiology faced were also present in Cambridge, such as a lack of space for students.[34] Therefore, the impact of physiology at Bart's is not straightforward and requires further investigation, but doubtless lacked the same influence on medical practice as the famous laboratory at Cambridge.

Examining language and physiology at Bart's and Addenbrooke's has revealed a completely different picture of the reception of laboratory science to that presented in Chapter 1. In contradiction to the indication of the reception of laboratory science revealed through terminology, bacteriological diagnostic methods were accepted more enthusiastically at Bart's than at Addenbrooke's. The following detailed studies of two physicians aim to reconcile why discourse was so different from practice at Bart's.

Samuel Gee

Samuel Gee began his professional career at Bart's in 1868, becoming a full physician in 1878. He was Physician to the Prince of Wales from 1901. With his education at University College School and a London medical school (UCL), together with his interest in the history of medicine and a position in the royal household, Gee can without doubt be classified as a gentleman physician.[35]

Gee has been portrayed as an advocate of medicine as an art. An initial glance at a plan for a lecture or an article within his private notebooks appears to support Lawrence's arguments about gentlemen physicians' beliefs in 'incommunicable knowledge':

> Begin with Hipp. Aph. i. 1. as epigraph: this introduces medicine as an Art. See back – 119 – science relates to knowing, art to doing. So innumerable are the terms (or rather things denoted by terms), so innumerable the complicns & modificns of these things, so imperfect an organ is language, that the science as written of med. can never contain all that may be & all that ought to be known. Hence much med. knowledge cannot be communicated: add to this that all men have to acquire the power of acquiring the necessary knowledge, i.e. have to cultivate their senses.[36]

Gee 'declined the honour of delivering the Harveian Oration at the Royal College of Physicians because he disapproved of Harvey's injunction to "search and study out the secrets of nature by way of experiment".[37] As well as apparently being opposed to laboratory science at the bedside (he was not opposed to physiology as a subject in its own right), it has been claimed that he was publicly opposed to specialization, proclaiming, 'There is a name that I hate, yea there are two names that my soul abhoreth, the name of specialist and the name of a consultant'.[38]

In contrast, evidence from the clinical case notes from Gee's wards shows that his patients received diagnostic assistance from the laboratory, therefore using the fruits of experimental science performed by a specialist in pathology. Table 2.1 shows that Gee's patients were diagnosed with Widal's tests in six out of seven surviving typhoid records in 1899. This shows an early and convincing acceptance of the worth of Widal's testing. For pulmonary tuberculosis and diphtheria, Gee's patients also increasingly receive bacteriological laboratory tests from 1894 and 1896 respectively. Indeed, as has been shown in Table 1.1, Gee was the second most prolific user of the specialist pathological laboratory at Bart's in 1895.

Table 2.1: The use of bacteriology in Samuel Gee's cases of pulmonary tuberculosis, typhoid fever and diphtheria at St Bartholomew's Hospital.

Year	Cases of typhoid fever	Widal's test	Cases of tuberculosis	Tubercle bacilli in sputum test	Cases of diphtheria	Bacteriological testing for diphtheria
1882	3	0	0	0	0	0
1883	6	0	6	0	0	0
1890	0	0	0	0	25	0
1891	0	0	0	0	10	0
1894	8	0	2	1	0	0
1896	0	0	0	0	20	14
1899	7	6	3	2	0	0

Source: Medical Registers, 1881–1920, St Bartholomew's Hospital Archives, MR 16/28, MR 16/36, MR 16/37, MR 16/40, MR 16/42, MR 16/48.

Within his private notebooks, Gee commented on some of the cases at Bart's, and it can be seen that he was actively using bacteriological diagnosis. These books include notes from other authors and his own aphorisms, together with some transcriptions of his case notes from Bart's. For a patient admitted on 3 May 1885, he wrote, 'A few days later several fragments of lung tissue in sputa, only one or two bacilli'.[39] For another patient, admitted on 12 May 1886, he wrote, 'There were no sputa at any time hence none to examine for bacilli'.[40] This shows how much Gee valued the test, as it was worth noting that he could not diagnose disease using bacteriology. He wrote, concerning a patient admitted on November 15th 1888, 'Sputa scanty, examined twice without result, but on Oct. 25 numerous tubercle bacilli were found'.[41] So, in this case it was worth testing again and again to achieve the confirmation which Gee expected: a bacteriological diagnosis of pulmonary tuberculosis. Finally, for a patient admitted on 16 September 1891, Gee noted 'tub bacilli not found'.[42] Despite not receiving bacteriological confirmation for all of the cases, Gee diagnosed them all as pulmonary tuberculosis, showing that he had reservations as to the accuracy of negative tests. Gee's notes also reveal the use of the specialist bacteriologist for a case of Cynanche cellularis from April 1896, for which he mentioned that Kanthack examined the cultivation.[43] As Gee's student, Andrewes recollected that in the 1880s he saw 'little application of bacteriology to the patients in the wards'.[44] Yet, Gee's notebooks show that bacteriology was used at Bart's in these years.

Although Andrewes had not realized Gee was using bacteriological diagnosis, this was not because Gee hid his use of the pathological laboratory. As well as carrying signs that Gee thought of medicine as an art, his published *Medical Lectures and Aphorisms*, which went through at least four editions, contains the following examples of his use of bacteriology: in an assessment of a case of chronic pituitous catarrh, read in front of the Royal College of Physicians, Gee announced that 'Tubercle Bacilli were not found'.[45] In discussing diagnosis of bronchitis and pulmonary tuberculosis within the same lecture, he stated: 'Even auscultation does not always enable us to distinguish these cases with perfect certainty, but since Koch's discovery of the tubercle bacillus we have a potent auxiliary to diagnosis in the microscopic examination of the sputa.'[46] In his *Aphorisms*, Gee said of cavernous bronchial breathing that if it was caused by 'tuberculous disease ... discovery of tubercle bacilli would confirm it'; and again concerning phthisis,

> Before the appearance of physical signs it is the presence of some or all of the following facts which leads to a diagnosis – haemoptysis, cough, loss of flesh and colour, slight rise of temperature, and hereditary liability to the disease; but above all, detection of tubercle bacilli in the sputa.[47]

He even advised his medical students to tell patients suspected of being 'phthisical' (literally wasting, referring to pulmonary tuberculosis) to 'Save your sputa to be tested (for bacilli)'.[48] Yet again, concerning typhoid accompanied by delirium, he considered that 'Testing the blood by cultivations of typhoid bacilli (Widal's test) [*sic*] must be employed as an additional means of diagnosis'.[49]

An admiration for laboratory sciences and specialization in general can be seen in some of his published writings. Before the Abernethian Society at Bart's in 1889, Gee compared himself with empirical physicians: 'I am ready to accept help from any source, from physiology and pharmacology, but also from mechanics, optics and similar sciences, from bacteriology, electricity and chemistry, nay very gratefully from cookery, upholstery, ironmongery, and indeed from any source.'[50]

He wrote a very similar comment privately, showing his conviction about the worth of other disciplines, though it may be that he saw the laboratory sciences as subordinate to medicine in grouping them with trades such as cookery, upholstery and ironmongery. Yet a hierarchy can be seen in this quotation, which starts with laboratory sciences followed by artisanal pursuits.[51] An admiration for laboratory science in general is demonstrated within Gee's public writings. In discussing the process of the lungs becoming emphysematous, again in a lecture to the Royal College of Physicians, he complained that 'unfortunately we can get no help from experimental pathology; we cannot set up progressive emphysema in the lower animals and mark its stages in the same manner as we can set up inflammation'.[52] These public proclamations about the use of bacteriology and laboratory sciences indicates that bacteriology was welcomed by the wider elite medical community, as Gee presented these arguments in front of audiences such as the Royal College of Physicians.

Gee may have advocated opposing ideas about the laboratory and specialists as a result of his multifarious personality. His colleague and friend, Wickham Legg, wrote that 'Gee's character was exceedingly complex, and it is very difficult to unravel. He had no zeal.'[53] Gee wrote publicly and privately about his dislike of dogmatism – the opinion that 'we can and do know'.[54] He considered that the fault in dogmatic sects lay 'in the small capacity of the human mind, unable to accommodate more than one thought at a time, and apt to see all things through the medium of this thought'.[55] He wrote privately, 'Dogmatism is bad, because abstract: let it cease to be abstract, & it becomes good.'[56] In spite of this scathing attack on dogmatism, Gee had the 'most kindly feeling' towards the 'Empiric sect', which was, apparently, although in his category of dogmatic sects, 'much less dogmatic' than his other examples.[57] Yet Gee is not a glowing example of an advocate of empiricism in the way in which he has been portrayed, as he argued: 'the empiric ... is the ant which gathers facts but does not transmute them ... his harvests are scanty. Well for him that all physicians are not empirics.'[58] Indeed, Legg considered Gee to be a sceptic.[59] Gee categorized scepticism in opposition to dogmatism, but he only examined sceptics in terms of therapeutic nihilism in his lecture on 'Sects'.[60]

Legg indicated that Gee was always ready to change his mind. Gee's lack of conviction about anything is illustrated by Legg's recollection which begins this chapter.[61] Certainly, Gee's aphoristic style of writing, revealed privately within his notebooks, and publicly within Horder's published collection of his aphorisms, was perfect for Gee's mutable mode of thought. Gee wrote privately: 'The advantage of the aphoristic way of writing is that no pretence of completeness is made; facts may be added without disarranging the structure. Science is taught thereby as a child learns a language, by disconnected scraps.'[62]

Gee doubted anything new. Twice he wrote in his notebooks about the unreliability of new facts: 'the majority of new "facts" are either not new or not true' and 'It is convenient to treat so-called new discoveries as Buonaparte treated his letters: wait for the year book, become acquainted with the new facts there, + most of them will turn out to be neither new nor true'.[63] This scepticism helps to explain why he may have been slightly reticent about the new experimental sciences infiltrating clinical medicine. Yet he appears to have considered many innovations and new ideas, not accepting any belief in particular. He did not wholly accept empiricism or experimental science. Legg's comment, 'Though he knew so much, yet his scepticism was very deep' sums up what is clear about

Gee – he was interested in everything, and accepted ideas and technologies on his own terms, always ready to change his mind.[64] Gee's complex personality has therefore resulted in contradictory published and private writings from which quotations can be found which exemplify both his opposition to and enthusiasm for the laboratory and specialists.

In addition to some of Gee's own discussions of his conservative gentlemanly practice, Gee's students' misinterpretations of his everyday work on the wards have contributed to his reactionary reputation. Andrewes's experience can perhaps be explained by Gee's belief in the knowledge which medical students should gain from the wards or from books. Gee wrote in his notebook, 'Always remember that the work in the Wards, so far as students are concerned, is simply the education of their senses. Pathology they can read at home.'[65] So, why did Gee feel that students could learn from their textbooks rather than directly from him about pathology? It is possible that Gee felt that the majority of students would not even have access to specialist pathologists. The majority would become general practitioners with a variety of roles, including Poor Law medical officers, life assurance assessments and working in factories and in schools.[66] However, at Guy's Hospital, the Clinical Research Association had been established in 1894, providing diagnosis by post to subscribers. A total of 4,000 doctors were subscribers, amounting to one in three on the Medical Register, as well as 150 organizations such as local authorities and small hospitals. However, the charge for tests was such that only wealthy practitioners and patients could have paid.[67] From 1896, Bart's pathological laboratory offered a similar service to readers of *St. Bartholomew's Hospital Journal*.[68] So, provision of specialist laboratory services was increasing. It is more likely that Gee felt elements of medical education were taught separately and that his role was to teach bedside manner and clinical examination. Bacteriology was taught in separate classes, first by surgical and medical staff such as Lockwood and Butlin, and from 1893 in three-month courses by Kanthack. By 1903, pathology was on the 'same footing as medicine, surgery and obstetrics' in the medical curriculum.[69] Also, case notes were used as a pedagogical tool, and Gee may have considered that the process of diagnosis for his patients demonstrated within his notes was a clear indication of his practice.

Another of Gee's students, Henry Dale, quoted him as saying, 'When you enter my wards your first duty is to forget all your physiology'. This has been repeated by Lawrence and by Kenneth D. Keele, who have used the same quote regarding physiology on the wards as evidence for the reluctance of physicians to accept the value of 'basic sciences' for medicine.[70] The presentation of Bart's by Gee's students has led Keele and subsequent authors to portray tensions in the acceptance of laboratory sciences into clinical practice at Bart's. However, the lecture in which this quotation appeared focused upon clinical research, not practice, and the discovery of treatments proven by experience. Therefore, although Dale presented Gee as being conservative in his method of choosing treatments and in the observational method by which he defined coeliac disease, it was not his everyday diagnostic practice which he discussed.[71] Dale studied at Bart's between 1900 and 1902. He had previously studied at Trinity College, Cambridge, and had undertaken a research studentship in the physiological laboratory at Cambridge for two years. His ability is indicated by his election as one of only ten undergraduates to be a member of the Natural Science Club, which at that time included Walter Morley Fletcher who became the Secretary of the Medical Research Committee (later the Medical Research Council) and Ernest Rutherford, who subsequently won the Nobel Prize for his work on radioactivity. Yet Dale did not secure further funding for research at Cambridge and decided to study medicine at Bart's. Living with his parents in London, he missed life in Cambridge, with its egalitarian atmosphere, and appears to have resented his experience at Bart's which he argued 'delayed my scientific development'.[72] Therefore, Gee's comment to Dale, to forget all his physiology, could have been a personal one about Dale's research career in relation to bedside practice, rather than a general opinion of laboratory science and medicine.

After medical school, Dale continued his research career at the Wellcome Physiological Research Laboratories where he became director in 1906. He became the head of biochemistry and pharmacology at the National Institute of Medical Research (NIMR) in 1914, and was appointed director of the whole institute from 1928–42.[73] The NIMR was part of the Medical Research Council (MRC), and in the 1920s and 1930s Dale was part of a conflict between the Royal Colleges of Physicians and Surgeons and the MRC. The Colleges complained that the work of the MRC was leading to a divorce between research and practice and worried that the government turned to the MRC rather than Colleges for major policy decisions. Walter Morley Fletcher argued that the Royal Colleges only represented leaders in practice, unlike the MRC which drew on a range of practitioners for the results of their investigations. Therefore, Dale could have had a strong agenda in portraying leaders of medical practice like Gee as negative about laboratory science in medicine, especially as his experience at Bart's was an unwelcome detour from his research career.[74] Having gone to some lengths to explain the contradictory representations of Samuel Gee, the next section applies the same queries to the publications and practice of a younger physician, Thomas Horder.

Thomas Horder

Horder's background was very different from that of Gee. His father was a draper, and he attended Swindon High School, not an elite school like Gee's. He was appointed as an assistant physician at Bart's in 1912, where he had studied medicine, and became a full physician in 1921. He certainly rose in status in his lifetime, becoming physician to monarchs and prime ministers: Physician-in-Ordinary to Edward, Prince of Wales; Physician-in-Ordinary to George VI; Extra Physician to Elizabeth II; and physician to Bonar Law, Ramsay MacDonald and Neville Chamberlain. He was made a baron in 1933. Horder aspired to be part of the England's elite. In addition to having an estate of 120 acres from 1924, he also corresponded with royalty about gardening.[75] He breakfasted regularly with Ramsay MacDonald during his premiership. However, as MacDonald was the illegitimate son of a crofter, Horder may have felt a close connection to another man who had successfully traversed the boundaries of class.[76] Being a gentleman physician at Bart's did not mean that a public school education and being a member of the upper middle to upper class was required, as gentlemanliness became more about behaviour than birth.[77]

Horder began working at Bart's when bacteriology was fairly well established as a discipline. Therefore his discussions addressed who should be undertaking diagnostic bacteriological tests and the extent to which their outcomes were valued. Horder has been represented as being opposed to specialization, and Lawrence has portrayed him as one of a group of elite physicians who were 'determined to integrate [the laboratory] into their preferred social order', an order where the ethos of holism meant that the clinician had to carry out all procedures himself.[78] Although Lawrence has subsequently slightly modified this view, he still considers that Horder wanted laboratory diagnosis to be carried out by the physician himself, although he was aware that time constraints were a problem. Apparently, Horder was optimistic that the more difficult tests like the Wassermann reaction for syphilis would be carried out in future by the clinician.[79]

Horder actually worked as a demonstrator in the specialist pathology laboratory at the beginning of his career.[80] In 1909 in a lecture to the Abernethian Society, which was reprinted in 1912 in the *St. Bartholomew's Hospital Journal*, Horder considered that the 'specialism' of bacteriology had produced 'excellent results'.[81] Although he thought clinical medicine should not be subordinated to the laboratory, he did not think it practical for the physician to carry out laboratory procedures:

> For, excellent though the principle is in theory, in practice it is quite impossible for the doctor to carry out by himself the many pathological investigations which are all essential for the proper diagnosis and treatment of his patients.[82]

The physician should, according to Horder, be familiar with the procedures, so they can know which to request from the pathologist,

> But to carry out the work himself is not possible; the hours of the day do not allow of it, and his technique cannot be kept sufficiently skilful to admit of accurate results. I realise this opinion is not held by everybody.[83]

He continued by criticizing a physician who made his patient wait for two hours for test results for parasites, and then returned the patient's stools for disposal as he apparently had no means of destroying them.[84]

In the following year, Horder emphasized the specialization of the pathological laboratory in his 1910 textbook, *Clinical Pathology in Practice*, in which he stated 'some care has been exercised to exclude descriptions of technique with which the pathologist, and not the practitioner, is chiefly concerned'.[85] The volume mainly discusses how practitioners should collect samples for specialist pathologists and how to interpret their reports, rather than the testing of the samples. Horder mentioned that students should learn how to do the tests in their degrees so that they could understand the methods and significance of these tests.[86] From the 1920s to the 1940s, Horder publicly supported specialization in general in medicine, not just within the laboratory. In 1933 he advised medical students in doubt of their career choice of the numerous worthwhile specialisms within which they could work, for example bacteriology, biochemistry, radiology and physiotherapy in addition to careers such as general practice. He also commented on how both the patient and the practitioner benefited from the act of specialization within the 'firm'.[87]

Unfortunately there are no case notes extant for Horder's inpatients who suffered from pulmonary tuberculosis, diphtheria or typhoid. However, an examination of his case notes in general shows that from twenty-six remaining files (in five of which he was assisting James Calvert), thirteen involved specialists in the pathological laboratory carrying out various tests for bacteria. Thirteen cases included other specialist services such as X-rays, blood counts, the electrical department, or specialist physicians and surgeons. Tests were carried out when patients were admitted for other illnesses. For example, in seven cases Wassermann tests for syphilis were ordered when syphilis was not the reason for admission. Only one of these tests was positive. Therefore, Horder was happy to use specialists, diagnosing patients away from the bedside. Indeed, from an opening address which Horder made in 1928 to the Association of British Pathologists, which was subsequently published for a wider audience in the *Lancet*, it is clear that Horder was only opposed to commercial laboratories:

> For the present unsatisfactory position [that of the doctor briefing the pathologist on the clinical features of the case in hand] in regard to this desideratum both parties are a little to blame. One fact which tends to obstruct this very necessary liaison is the existence of laboratories in which the personal element as between doctor and pathologist is quite eliminated. Materials are dumped in these places much as coals are dumped at our houses. I supposed such places are necessary; anyway they seem to have come to stay. I do not patronise them myself, because I am fortunate enough to be in frequent daily touch with colleagues working in pathological laboratories.[88]

What was important to Horder was communication between the clinician and the specialist. This is demonstrated in this correspondence between himself and the radiologist at Bart's in 1920. Having received an X-ray report requested from the radiologist, which stated 'nothing abnormal', Horder responded:

> Dear Stone
>> Is not this mediastinal shadow (ant. post.) rather wider than usual (aet 45).?
>> And is the heart shadow normal?
>> Yrs,
>> Thomas Horder

Dudley Stone replied,

>> Yes.
>> No.
>> I did not think the appearance of either to be other than a normal abnormality – not pathological. The rounded ventricular shadow suggests a loss of muscular tone.[89]

It is easy to see how Horder's position regarding specialism in medicine has been misunderstood. He regularly makes comments regarding the problem of equipment distracting from clinical diagnosis, for example,

> when I am faced with a mass of data resulting from the exploitation of instruments of precision, I ask the patient, as soon as I can isolate him from the laboratory equipment, 'Where does it hurt you?' and then listen carefully to what he has to say.[90]

In contrast, although he advocated the importance of bedside medicine, he also publicly stressed opinions such as '*[n]ever neglect to confirm the diagnosis of phthisis by the demonstration of tubercle bacilli in the sputa, however "classical" the signs and symptoms may be*', and in cases of bronchial catarrh being possibly tuberculous, '[t]he final court of appeal must always be the search for tubercle bacilli in the sputa: three successive negative examinations of material *which is properly chosen* may be considered to exclude phthisis'.[91] Yet, the use of the laboratory was not straightforward. Horder discussed cases of typhoid where even though all of the clinical symptoms and signs were clearly present, there could be a negative bacteriological test.[92] Just like bedside medicine, the laboratory was not infallible – there may be false negatives in bacteriological tests.[93] This same problem of avoiding error in the laboratory is still a concern in the present day, with a recent call for communication, and in difficult cases, for brainstorming sessions by both the pathologist and the clinician.[94]

Horder clearly had no reservations in using the hospital laboratory. Lawrence has made much of 'Horder's box': his own portable laboratory for his private patients.[95] But when he treated patients in the hospital, Horder was happy to use the services of the pathological department. Additionally, Horder had worked in the pathological laboratory and had written a textbook on the subject, so was much more highly skilled in bacteriology than the average physician and could conduct tests himself. Horder was willing to accept a certain degree of specialization as long as each specialist could communicate, in order to form a full picture of the patient's clinical, bacteriological and radiological signs and symptoms.

Integrating the Laboratory into an Elite London Hospital

Terrie Romano has shown that a gentleman could partake in new laboratory sciences such as experimental physiology, and Peter Mandler has remarked that the upper classes in general were modern in their outlook.[96] Therefore, it appears that being a gentleman did not equate with opposition to the new techniques and practices of the laboratory and specialization. Certainly, Chapter 1 demonstrated that William Church, educated at Harrow and president of the Royal College of Physicians, was one of the main campaigners for a pathologist to be appointed at Bart's. He also argued for the pathologist to receive higher pay, and to have enough time for original research.

There was no distinguishable difference between the use of bacteriological diagnostic techniques by those who had been to public schools and those who had not. This is in line with David Edgerton's argument that Britain was not as technologically backward as declinists believe. He protests against the most popular explanation of the so-called decline: the idea that education was anti-technology and anti-scientific.[97] Indeed, a classical education did not affect bacteriological diagnostic practices. Physicians educated at the Clarendon schools (the top nine public schools in England) used bacteriology for 100 per cent of the diphtheria cases they saw in 1896. The public school system has been portrayed as detached from the 'modern world', with science almost absent from curricula, and Oxbridge being opposed to industry.[98] Yet attending public schools did not necessarily mean that physicians were not educated in science, as its place in the curriculum was increasing.[99] Three of the nineteen physicians who worked at Bart's between 1880 and 1920 attended either Harrow or Rugby.[100] Science was increasingly taught at Harrow from 1805, and at Rugby, science was taught from 1851, with a special school for the subject being built in 1856. In 1862, Royal Commissioners found a choice of either modern languages or natural sciences at Rugby, and a physical science lecture room and laboratory costing over £1,000, with most of the teaching being on chemistry and electricity.[101] So an elite public school education certainly did not mean a lack of exposure to science and the laboratory.

Examining Gee and Horder's arguments in light of perceptions of their professional identity rather than their views of practice reveals a more complicated picture. Recently, Lorraine Daston and H. Otto Sibum have discussed the possibility of scientists, and people in general, having different personae depending on the context.[102] However, the traditional idea of double lives considers people have secrets to hide.[103] This chapter has challenged views of double lives which state that different behaviour is displayed within public and private arenas. Gee presented a mixture of discourses in both of these provinces. He considered that the worlds of bedside medicine and laboratory science were compatible with each other, even though he sometimes portrayed himself as a clinician who considered medicine to be only an art.

Yet there was a significant reason for the physicians to be cautious about the impact of bacteriology and the 'ancillary sciences', in order to defend their status. The low status of the bacteriologist, discussed in Chapter 1, shows why the physicians generally did not feel as though they were vulnerable to a professional challenge when they discussed the value of bacteriology in public, with the profession of a pathologist linked with artisanal occupations. However, diagnosis was perhaps the most important technique for a physician.[104] Gee remarked in his *Medical Lectures and Aphorisms*, that 'diagnosis' was

> a most important affair for the empiric, inasmuch as his line of treatment depends thereon. So that the empiric would almost agree with him who exclaimed that the first part of the treatment is diagnosis, and the second diagnosis, and the third diagnosis. After the diagnosis comes the fourth and therapeutic stage of the method, in which lies the essence of empiricism.[105]

Although he was not uncritical of the practice of the empiric, this was the medical sect he had most closely aligned himself to. Horder and his colleague, Walter Langdon-Brown, considered diagnosis was the 'key to medicine'.[106] Along with another colleague, A. E. Gow, Horder argued in 1952 that diagnosis was the 'be-all and the end-all' of medicine.[107] Therefore, bacteriology was quite a threat to physicians, as their major clinical role was being challenged.[108] Gee privately wrote: 'Medicine is independent of the so-called ancillary sciences tho' gladly receiving aid from them. Medicine can rest upon its own foundations. Clearly the perfection of physiology, healthy and diseased, + therapeutics would be the annihilation of medicine.'[109]

Popular understanding of diagnosis and medicine generally, including the questioning of expert knowledge, has been discussed with reference to X-rays. The X-ray became popular with the public overnight – new quality levels of printing in the late nineteenth century in magazines allowed the public to see the same images as the medical profession. Joel Howell notes that in the late 1890s, it was generally regarded that lay people 'could derive some meaning from an X-ray image.' In one way this was useful as it could be used to persuade a patient of their diagnosis. On the other hand, who had authority over the X-ray? Records of hospitals in Pennsylvania from 1905 and 1913 show that although there was a charge to the patient for the X-ray, the plate itself would remain in the hospitals' possession.[110] In 1907 and 1913 it was discussed in journals that a patient had no more right to an X-ray plate than to a pathological slide or urinary sediment, as all need expert knowledge for interpretation.[111] X-rays changed from being part of popular culture to being 'mystified'.[112]

In the 1920s and 1930s, Horder seems to have been worried about the profession of the clinician as a result of patients' perceptions. In his experience, his patients were growing increasingly convinced of the power of machines and experts.[113] In 1924, he remarked that lay people were treating doctors as though they had 'no expert knowledge whatever', putting forward ideas such as possible causes of cancer.[114] In the 1930s, patients were coming to see him armed with X-rays and ideas about the reasons for their illnesses.[115] He defended his profession, stressing the importance of 'human judgment'.[116] The effect of the depression may have affected Horder's views regarding laboratory diagnosis in the 1930s. He commented on how patients were coming to him having already had diagnostic tests to save time and money. One patient would not allow Horder to even examine him, saying, 'please give me something for my headache: I don't want a diagnosis, it costs too much'.[117] According to Horder's publications and lectures, lay knowledge of disease appears to have been increasing in the 1930s. In 1937, Horder introduced a collection of addresses which he published by saying:

> I have consigned two of the addresses which contain a certain amount of technical matter to the end of the series. So much of our medical jargon is known to the public these days that in doing this I have perhaps shown an unnecessary concern for the reader.[118]

Ideas of the increasing gap between the knowledge of doctor and patient due to laboratory medicine seem to be exaggerated.[119] Horder was not alone in his experience, as Christopher Crenner has discussed the comments which American patients made to the elite Boston practitioner and specialist in diagnosis, Richard Cabot. Cabot did not serve elite patients, rather skilled labourers, small businessmen and a few people from the professional classes, but still he faced patients disputing his diagnoses in the early twentieth century.[120]

Both economic problems and the expansion of the allopathic medical marketplace threatened the position of the general practitioner in the 1930s. Although Horder may be considered to have been an elite consulting physician rather than a general practitioner, his experiences with private patients can be compared to that of GPs. According to Anne Digby, GPs had to compete for survival amongst specialists after 1920.[121] In the 1890s, specialization within the medical profession had not caused so many problems as there was generally a referral system in place, but this was not the case by the 1930s.[122] The Political and Economic Planning Report of 1937, which surveyed the existing health services in Great Britain suggesting future developments, made repeated comments about the importance of the GP as a specialist in diagnosis.[123] It was argued that the patient should see the GP first and he would decide if another specialist should be used.[124] It was repeatedly stated that the private practitioner was 'losing ground' and needed to be more modern in outlook.[125] In particular, the public were apparently 'hypnotised by the word "specialist"' and the report warned against the tendency of people running straight to specialists without consulting their GP.[126] Apparently it was necessary to restore public confidence in GPs – the reluctance to see one was according to the report not always 'unjustified'. However, the report contrasted this negative view of GPs with patients who considered them to have gifts of 'magic'.[127] The affordability issue was also mentioned in the report with regards to the future of the private practitioner – he must 'find a more efficient economic basis for his work'.[128] Digby notes that not only other doctors, but also nurses, midwives and health centres dealing with problems such as tuberculosis, foot care and child welfare were alternatives to seeing the general practitioner, and these services were often cheaper.[129] However, the financial problems which threatened GPs in the 1930s were unfounded.[130] Despite this, the atmosphere of this decade appears to have changed Horder's attitude towards the laboratory.[131]

Wider political concerns also affected Horder in the 1930s. A strong theme in his writings during that decade is individualism – that the doctor should not be swept away by mechanization and should treat each patient individually.[132] This idea appears to have been influenced by the politics of the 1930s, and the fear that what was happening in the USSR and Germany would remove individualism. In 1936 he considered that the fate of the world depended on whether Northern and Western Europe and North America managed to retain their individuality. He related this to medicine – it was important to treat the individual rather than the masses – and he stressed that the political leaning of the medical profession should be towards liberalism and that political ideology needed a 're-birth'.[133] By the 1930s, Lawrence's representation of Horder is much more accurate. Professional status was threatened if patients went to radiographers and bacteriologists to acquire cheaper diagnoses, if they bypassed general practitioners to see medical specialists, and if patients' increasing faith in machines outweighed their faith in the expertise of the physician and bedside medicine.

Conclusion

This chapter has shown that the acceptance of bacteriological diagnosis at Bart's is not straightforward. Examining language has demonstrated that although quantifiable measurements such as temperature were influenced by the discourse of experimental science at Bart's, a transition in discourse was much slower for the description of subjective changes within the body. This indicates a difference in mentality between Bart's and Addenbrooke's with regard to experimental physiology, and perhaps indicates that it was the restriction in funding rather than the attitude of the doctors which resulted in a more gradual use of bacteriological diagnosis at Addenbrooke's. However, as Risse and Warner have discussed, it is important to analyse both physicians' behaviour and their rhetoric in order to understand the meaning of their behaviour.[134] Evidence from Chapter 1 cannot be ignored, as in every way Bart's physicians used the new bacteriological laboratory and its specialists' techniques more routinely than at Addenbrooke's. This may be due to the haphazard completion of many of the Addenbrooke's case notes, but it is the only evidence that this study can be based upon. The short biographies of Gee and Horder have accounted for the representations in the historiography of Bart's, illustrating that they were not opposed to laboratory science between the 1880s and the 1920s.

By considering how the public interpreted expertise, specialization and the laboratory, this chapter links to the next four chapters, which examine how bacteriological knowledge and evidence was received in society. The main aim of Chapters 3 to 6 is to investigate lay knowledge and use of the findings of bacteriologists, but in so doing, opinions of the medical profession will continue to be discussed, as well as those of journalists, lawyers, activists and politicians.

3 ANTHRAX IN BRADFORD: UNDERSTANDING DEADLY DISEASE IN THE WORKPLACE, 1880–1905

He thought the whole thing a huge Bacterial bubble which would not be long before it burst.

On Edward Tibbits,
Bradford Medico-Chirurgical Society Minute Book, 4 April 1882[1]

Woolsorters' disease was a major concern in late nineteenth- and early twentieth-century Bradford. Woolworkers in Bradford are recorded to have first suffered from the disease between 1838 and 1847 and then with increasing publicity from the late 1870s.[2] In the late 1870s and early 1880s, the disease was determined to be a pulmonary form of anthrax that was introduced into the wool trade through imported wool from the East. By 1860, imports of wool had overtaken home-grown wool.[3] These imports may have led to the disproportionate dread of anthrax as a result of its association with the dangers of the East, similar to the fears of Asiatic cholera which came in waves from the East during the nineteenth century.[4] Despite the rarity of the disease, anthrax's frightening and sudden attack on the body may also have led to this fear.

In its respiratory form, anthrax is shocking, as symptoms develop so quickly that it can kill before the disease has been diagnosed. Inhalation of spores was the form of transmission which killed most of the wool workers who succumbed to anthrax in Bradford.[5] Cutaneous anthrax is also a very dramatic but visible form of the disease. Presenting with black pustules, it usually requires a scratch in the skin through which the bacilli can enter. Gastrointestinal anthrax occurs when the bacilli or spores are ingested. All forms of anthrax can develop into a systemic infection, with the possibility that victims can appear to be fine one day, develop a slight headache, and die the next day. In between, the victim can suffer malaise, headache, fever, gastrointestinal problems, possibly red boils which turn black, a cough, seizures and organ failure. In order to survive, *Bacillus anthracis* has to infect the body and kill it, causing external bleeding in order to release the bacteria, in the form of spores, back into the atmosphere. Organs, especially the spleen, are 'almost liquefied'. Two lethal toxins are released by the bacteria within the victim's body to ensure death. Untreated, the cutaneous form has a 20 per cent fatality rate, the gastrointestinal form has a 50 per cent rate and, if struck by the inhaled form, the victim only has a 10 per cent chance of recovery.[6]

Robert Koch's demonstration of the life cycle of the anthrax bacillus in 1876 has often been cited as the beginning of the bacteriological era and as a 'model' for future bacteriological research.[7] From 1873, the rural physician studied the disease which infected both animals and humans where he lived and worked in Wöllstein. Influenced by emerging germ theories in France and Germany, Koch studied the bacilli, observing their behaviour in different areas of a droplet, becoming longer and granular towards the edge and forming spores at the very edge. If the spores were warmed again in aqueous humour, they returned to their bacilli form. Koch's understanding of the bacilli's life cycle revealed how they could survive for years in fields after animals were buried, before infecting new livestock.[8]

Yet Koch was not the first to discover the bacillus, nor to experiment with it. In Koch's work and that of his predecessors, the study of the anthrax bacillus was the most developed of all microorganisms at this time, the study being aided by this microorganism's large size and its receptivity to staining. In Wipperfürth, near Cologne, Franz Pollender discovered the organism in 1849, but was unable to define its exact link to the disease. In France, Pierre Rayer inoculated a sheep with blood taken from an animal which died of anthrax in 1850 and with a microscope saw the organisms in the inoculated sheep's blood. In 1863, Casimir Davaine demonstrated that anthrax could be contracted by sheep, horses, cattle, guinea pigs and mice, and that a specific microorganism could transmit a specific disease. Davaine had been Rayer's student and he claimed priority in spotting the small elongated organisms – 'bacteridia' – in the blood in 1850. In what is now Estonia, Friedrich Brauell experimented on animals using blood from a man who had died from cutaneous anthrax. The French veterinarian, Henri Delafond, speculated in 1860 that the rod-shaped organisms may arise from spores and were associated with anthrax in some way. In Massachusetts, these organisms were seen by physicians examining the blood of victims of an outbreak in a factory in 1868.[9]

Anthrax is a key disease when examining the reception of the concept of pathogenic bacteria, as the study of this particular bacillus was well underway before bacteriology emerged as a discrete discipline. When Koch carried out his anthrax research, there were only a very small number of experimental pathology laboratories in Britain.[10] The anthrax bacillus infected the woolsorters of Bradford in the late 1870s, and the Yorkshire town thus provides an ideal case study for analysis of working-class engagement with a new scientific theory. Indeed, Peter Bartrip has argued that between 1878 and 1881, the town was Britain's 'unlikely setting' for the 'leading edge of medical research and a testing ground for the emergent germ theory of disease'.[11]

During the late 1870s, journalists and local doctors campaigned for the discovery of a way to control woolsorters' disease in woollen mills. Dr John Henry Bell of Bradford, a local physician who already had a reputation for investigating occupational disease amongst miners, became involved. Using the laboratory, he demonstrated that the disease was anthrax, and certified the death of a woolsorter as an employer's responsibility in 1880, thereby involving the police in a coroner's inquest. This resulted in inquiries by the Bradford Medico-Chirurgical Society and by the Local Government Board (the government department overseeing local government, public health and the Poor Law, hereafter LGB) and increasing local voluntary regulations in Bradford during the 1880s and 1890s.[12]

This increasing publicity resulted in anthrax being recognized as a notifiable disease within the Factory and Workshops Act of 1895.[13] The iconic status of anthrax as the first bacterial disease included in the Workmen's Compensation Act was cemented in 1906. Research and regulations continued at a local and a national level. The Bradford and District Anthrax Investigation Board (originally called the Anthrax Committee) was established in 1905 by various parties in Bradford and the surrounding area. In 1913 the Home Office appointed a committee to experiment with disinfection which resulted in the establishment of a disinfection station in Liverpool following the Anthrax Prevention Act of 1919.[14] Chapter 4 will explore how investigation of anthrax continued at an international level through the initiative of the International Labour Organization.

Recently, several scholars have argued that bacteriology was not particularly influential in the campaign to prevent anthrax, or in the decline in cases of the disease. This chapter tests the ideas of four historians in particular. James Stark has proposed the former argument, suggesting that measures were allied to practices of the sanitarian movement which preceded bacteriology, focusing on the environment.[15] Peter Bartrip has argued that 'medical advances in prevention and treatment' and regulatory measures were less likely than an improved standard of living to have reduced the incidence of industrial anthrax, allying this transition to Thomas McKeown's argument that resilience to infections was more important than innovations in medical science in reducing mortality rates.[16] Ian Mortimer and Joseph Melling have persuasively suggested that the decline in trade in the 1930s was more effective in reducing anthrax than the 'scientific war on the bacillus'.[17] This chapter does not seek to dispute these claims and does not examine the decline in mortality in detail, nor the rules and regulations within the workplace. It seeks to regain a place for bacilli in the story of anthrax prevention by asking how knowledge of bacteriology influenced campaigns to prevent the disease.

The present chapter focuses on relations between workers, the medical profession, lawyers, the public and the press during the incidences of anthrax from 1875 to 1904–5. As Melling has summarized, historians have recently approached occupational diseases by considering that recognition and action was driven by communities of workers more than by the introduction of new medical technologies.[18] How did this play out when anthrax struck Bradford at the same time as the aetiology of the disease was being defined in the laboratory? Melling argues that twentieth-century British activism relied on networks of medical practitioners, experts and campaigners from both inside and outside of the workplace.[19] How did people without medical training perceive early bacteriology, how did they acquire and utilize their knowledge, and how did conflicts within and between different groups affect their actions regarding local incidences of anthrax? In order to understand the perceptions of employees, journalists and lawyers, it is important to examine how the medical profession presented their ideas in public, and any conflicts between doctors. Therefore, this chapter's purpose is to understand how bacteriological knowledge was used by various historical actors during the very early years of the discipline, rather than aiming to be a complete study of the disease in Bradford.[20]

Ending with the events of 1904–5, this chapter concludes at a sea change in the history of anthrax. In 1904, Thomas Legge, the first Medical Inspector of Factories, visited a laboratory in Italy to learn about Achille Sclavo's serum which had been developed in 1895 in response to the high number of human deaths from anthrax in the country: 7,308 between 1890 and 1904.[21] Before 1904, the serum was not available for export as it could not be produced in a high enough quantity and was subject to rapid degradation.[22] This therapy was subsequently used in Britain to treat anthrax sufferers, resulting in a 50 per cent decline in mortality from 1906. However, it was very expensive and fatalities still occurred even when large doses were used, though with a quicker diagnosis the serum was more effective.[23] In addition, early successful claims for compensation also occurred at this time. This chapter examines the earlier period when the local doctors and workers were more informally involved in fighting anthrax. Historians have already discussed in detail the actors and organizations involved after this date, including central government; it is the initial understanding of different groups and their use of bacteriology which is the aim of this chapter, rather than the analysis of resulting legislation.[24]

The incidences of anthrax investigated within this chapter are split into two distinct periods: before 1889 and between 1899 and 1904. Table 3.1 illustrates that the deaths from anthrax in Bradford were apparently divided by an eight-year reprieve between 1892 and 1899 when there were no recorded fatalities in the Medical Officer of Health's (MOH) reports. This is also demonstrated by the gap in the letters to the press from workers and the public (see Table 3.2).

However, the structure of the chapter is a thematic rather than a chronological examination of these two periods. In order to understand the use of bacteriological knowledge in Bradford, the following issues will be examined in this chapter: the political context of Bradford; the presence of anthrax within the wool trade in the late nineteenth century; the response of the medical profession and the workers; dissemination of information in the press; and how lawyers utilized knowledge of bacteriology in compensation claims. The agendas of these actors and the conflicts within and between these different occupational groups will be discussed throughout. However, first the reasons why a relatively rare disease has attracted so much attention from scholars is explored.

Table 3.1: Number of deaths from woolsorters' disease/anthrax per year and recorded bacteriological tests in Bradford and occupation of victims where available.

Year	Deaths	Records of bacteriological tests or other details in relation to the disease	Occupations
1875	0 deaths		
1876	No mention		
1877	2 deaths		
1878	3 deaths		
1879	2 deaths		
1880	6 deaths		
1881	3 deaths	1 external case	1 carder, 1 back-washer, 1 machine woolcomber
1882	1 death		
1883	2 deaths		
1884	3 deaths		2 woolsorters, 1 card minder
1885	1 death		1 woolsorter
1886	2 deaths		1 woolsorter, 1 wife of a woolsorter
1887	2 deaths		2 woolsorters
1888	3 deaths		
1889	0 deaths		
1890	1 death		
1891	1 death		
1892	No mention		
1893	No mention		
1894	No mention		
1895	No mention		
1896	No mention		
1897		3 bacteriological examinations for anthrax	
1898		5 bacteriological examinations for anthrax	
1899	No mention		
1900	2 deaths		1 card jobber, 1 woolcomber
1901	3 deaths		1 foreman woolsorter, 1 card feeder, 1 card jobber

Year	Deaths	Records of bacteriological tests or other details in relation to the disease	Occupations
1902	1 death	2 cases malignant pustule recovered 12 bacteriological tests, 3 positive	1 woolsorter
1903	2 deaths	11 other cases, 7 bacteriological tests, 2 positive	All engaged in different stages of wool process
1904	1 death	3 other cases, 4 bacteriological tests, 0 positive	1 woolpacker Either washers or sorters

Note: When there was no special paragraph on the disease in the report, information regarding the occupation was not available.

Source: Borough of Bradford, *Reports on the Health of Bradford by the Medical Officer of Health, 1875–1904* (Borough of Bradford: Bradford 1876–1905).

Historicizing Anthrax

Anthrax has attracted the attention of historians of human and animal medicine, anthropologists, occupational health specialists and physicians with an interest in history. There are several accounts of the story of anthrax in Bradford, including book chapters within volumes on the disease, and three doctoral theses which either concentrate on Bradford or include comparisons with anthrax in the city as part of a study of anthrax in Kidderminster.[25] In addition to attention on anthrax in Bradford, several other topics have especially attracted historians. Anthrax has been associated with the plagues of Egypt and even considered as an alternative cause of the Black Death.[26] Codell Carter discusses the nationalistic rivalry between Pasteur and Koch in the late 1870s and early 1880s when Koch claimed that Pasteur falsified his anthrax vaccination experiments.[27] Bruno Latour explains these dramatic anthrax vaccination demonstrations in the theatrical Pouilly-Le-Fort public laboratory.[28] Recent biographies of the disease examine the history of anthrax in the light of the bioterrorist attacks on the American postal system.[29] Worries of anthrax and biowarfare also led to Jeanne Guillemin's book examining whether a USSR outbreak in 1979 was simply from infected meat, or the result of an accident in a military compound.[30]

Anthrax also seems to have disproportionately grabbed attention in the nineteenth century. Mitchell Brothers, manufacturers of woollen products in Bradford, wrote a letter to the press in March 1878 complaining that the idea was 'gaining ground that the occupation of mohair and alpaca sorting is more destructive to human life than almost any other occupation' even though they could prove otherwise.[31] Legge considered that the public would be surprised that the case fatality rate was only 25 per cent, as they thought recovery was almost impossible.[32] Historians Mortimer and Melling ask why the British state paid so much attention to the control of a disease which posed a 'limited health risk to a regional workforce'. They state that the 'extraordinary attention devoted to the disease' is as a result of Bell's identification of the disease as an industrial injury and its subsequent definition as an industrial accident, in addition to the dramatic appearance and sudden deaths of anthrax sufferers.[33] This chapter proposes that it was the location of Bradford, the conflicts between the various actors, and the timing of the anthrax cases which led to dramatic representations and significant publicity regarding the disease. Before examining anthrax in Bradford, the political context of the town will be discussed.

Political Bradford

The socialist cause was advanced in Bradford, and as Tim Carter has noted, may be the reason why there was such a vigorous campaign against anthrax in the town.[34] The Independent Labour Party founded its national organization in Bradford in January 1893. Keith Laybourn and David James, historians of this movement, argue that Bradford was the clear choice for this meeting as it reflected the area's 'move towards political independence for working men and Socialism in the early 1890s'.[35] The party not only supported trade unions but its clubs organized classes in a variety of themes, from singing and dancing to political economy, with one of the purposes being to educate the working class so they could successfully agitate for improved conditions. This movement began in the 1870s, with the depression in Bradford's wool trade resulting in changes in employment including the introduction of new machines, deskilling, wage reductions and unemployment, in contrast to the benefits of the trade which the workers had enjoyed in the buoyant 1860s. This dramatic change led to militancy in Bradford, including socialism and trade unionism.[36] A problem in the wool trade was a problem for Bradford as a whole and bound to gain local support; in 1850 the trade employed over half of the town's residents, and even during the economic depression of the late nineteenth century, it was still the town's largest employer.[37] Despite woolsorting being a respected job, which required experience and knowledge of the ways in which wool was classified for its quality, sorters found their work increasingly insecure, with employers looking to employ part-time sorters and to increase worker productivity from the 1880s.[38]

In 1880, the woolsorters began striking in order to express their concerns about woolsorters' disease. A strike calling for precautions to prevent disease at Harden Mills was discussed in the *Bradford Observer* on 8 May 1880:

> The majority of the woolsorters employed by Messrs S. Watmuff and Co., manufacturers, of Harden, who refused to continue in their employment owing to their objection to sort Van mohair which had not been purified in the manner desired, have not yet returned to their work, although a few of them did so this morning.[39]

This was in response to the death of two of the twenty sorters in April as a result of the pulmonary form of anthrax, known as woolsorters' disease. Two more sorters had malignant pustules, the cutaneous form of anthrax. Two combers at the factory also suffered, one from the internal form and one from malignant pustule. They had all worked with Van mohair, a wool from goats which was imported from Turkey.[40] The necessity of the workers' strike was met with sadness in the *Bradford Observer*, which reported that although a 'more justifiable strike than that of the woolsorters at Harden Mills could not be imagined ... such a mode of enforcing the proper action is unsatisfactory and incomplete, and should assuredly not be necessary'.[41] The newspaper reveals that this was not the first strike in Bradford regarding anthrax. In June 1880 it was noted, with reference to Harden Mills, that

> A somewhat similar case was reported last year, with a very dramatic ending. A firm which had been in the habit of steaming its dangerous wool became too busy to afford the time. The sorters refused to work the unsteamed material, and their places were filled by others. Within a few days, three out of twenty new sorters were dead.[42]

The *Bradford Observer* was the leading newspaper in the district, and was printed by an activist publisher and advocate for the Liberal Party, William Byles. Among other campaigns, the newspaper advocated social reform. Despite an early allegiance with the Tories in opposition to the Liberal-Party-supporting factory owners, the working class had turned to the Liberal Party, which supported suffrage. However, Byles aimed to present a balanced point of view in the newspaper.[43] Byles presumably sympathized with the working class within his newspaper: he eventually bowed out of party politics in despair as he wanted to adapt the views of the local Liberal Party to be more receptive to working-class demands. He was one of the few middle-class supporters of a strike in 1891.[44] Yet, James Stark has argued that the newspaper was selective in its campaigns during the 1870s, and did not support trade unions. In spite of this, he believes that campaigning on behalf of woolsorters within the newspaper was important in order to ensure the continued success of the major local industry.[45]

Striking in Bradford was certainly not novel; there was mass striking in 1825, for example, when the woolcombers demanded higher wages. Their employers did not allow them back to work for five months, and finally there was a ruling in favour of the employers. The woolcombers accepted defeat and returned to work with the same pay and hours that had caused them to strike. However, they formed the Woolcombers' Protective Society.[46] Hand combers were central in Chartist riots in Bradford in 1848.[47] Therefore, it was probably not too remarkable that some of the workers decided to strike for safer conditions in 1880. It is perhaps more surprising that striking was not more widespread. This may have been because of efforts from the middle classes to incorporate working-class leaders into the local Liberal Party and to improve the borough environment, or because of the increasing insecurity of employment. A subsequent, unsuccessful five-month strike at a mill occurred in 1891, but this time the focus was on opposing a wage cut.[48] Despite the burgeoning political activism in the town, there were not as many strikes regarding anthrax as may have been expected. Yet the presence of various societies and unions, including the Woolcombers' Protective Society, the Woolsorters' Society, which was a mutual relief society, and the National Woolsorters' Union, resulted in continued organized pressure.[49] The following section examines the extent of deaths from anthrax in Bradford before analysing the reactions and interactions of different communities. Although this chapter is especially intended to analyse lay responses to bacteriology, it is essential to first examine the activities of the local medical profession who collaborated to investigate the transmission and the course of the disease.

Anthrax in Bradford

A letter written to the editor of the *Bradford Observer* in 1878 is frequently cited in the literature on anthrax:[50]

> Sir – Within a month three woolsorters have died from blood poisoning contracted in the same shed of the same factory in Manchester Road. Occasional deaths from blood poisoning, through the handling of some foreign wools, may perhaps be inevitable; but the recurrence within so short a time of three such cases seems to suggest either that the wool itself, or the shed in which it was sorted, is especially unwholesome. May I hope for your help to induce manufacturers to remedy all that is remediable in this deadly trade? No doubt the death of a woolsorter means to the manufacturer merely the loss of an easily replaceable hand, but to others it means the loss, never to be replaced, of son, husband, or father. That I have just come from one such desolate home must be my excuse for troubling you with this letter – I am, &c.   X.
>
> Low Moor, Feb. 25[51]

It has been cited as the beginning of the press campaign, and it is the first cutting pasted into the main primary source for this chapter, a collection of news cuttings made by Bell.[52] Stark has speculated that this letter may have been written by a local doctor, William Rawson who lived at Low Moor, as it echoes Bell's views which were not yet published but had just been presented at the Bradford Medico-Chirurgical Society.[53] This letter was later referred to in the *Lancet* by Frederick Eurich, the City of Bradford bacteriologist who worked for the Anthrax Investigation Board for Bradford and District and subsequently for the Home Office Inquiry.[54] Eurich claimed that, since the date of that letter, the 'interest of the Bradford public in the anthrax question has never waned'.[55]

The letter from Low Moor states that deaths from blood poisoning are expected in the wool trade, so it would be easy to assume that these three deaths from woolsorters' disease were only some of those experienced by Bradford workers in 1878. A subsequent letter was sent by 'Chirurgicus' to the newspaper, arguing that woolsorters' disease

> affects the health of a large section of the working population of this town and neigh-
> bourhood, and requires prompt consideration on the part of those whose duty it is to
> deal with all matters pertaining to the health of the district.[56]

However, an examination of the MOH reports suggests that the dangers were limited. The three deaths mentioned by the resident at Low Moor constituted the total deaths from woolsorters' disease in Bradford in 1878. Additionally, the deaths are also shown in Table 3.1 not to be exclusively amongst woolsorters, but among wool workers more widely, especially in 1881.

Nationally, 261 cases of anthrax in factories and workshops were reported to the Home Office between 1899 and 1904, 67 proving to be fatal.[57] Bradford accounted for a third of British deaths, which was very high for one town. How-ever, deaths documented in reports for Bradford presumably did not include those for the surrounding areas in the wool trade, such as those recorded in the press in Keighley. Yet this was still a low number of deaths over a six-year period in com-parison to other infectious diseases. For example, in Bradford in 1880, 189 people died from scarlet fever, 18 from diphtheria and 66 from measles, compared to 6 deaths from anthrax.[58] As previously mentioned, there were complaints about the exaggerated number of deaths. In the press in March 1878, Mitchell Brothers complained about the representation of the trade of woolsorting as dangerous, and that it was untrue that deaths had recently been increasing.[59] Harris But-terfield, the MOH, also wrote in March 1878 to explain that woolsorting was a healthy trade based on his statistics. However, Butterfield was discussing deaths in the trade as a whole when he examined the health of the woolsorters.[60] The *Bradford Observer* concluded in the same month that it did not matter how many deaths there were: 'The point at issue is not what statistics say as to woolsorting when compared with other employments; but whether there is not an obvious and admitted cause of death amongst woolsorters which does not act on the gen-eral community'.[61] Also, in the *Bradford Observer* in May 1880 it was announced that deaths had previously been entered under other causes, and Bartrip argues that even after this time, cases of the disease continued to be misdiagnosed.[62] So Table 3.1 may well understate the number of deaths from the disease.

'Bacteriomania?' The Bradford Medical Profession and Anthrax

In early March 1878, an inquest determined that the death of a woolsorter was caused by the inhalation of 'vitiated' or 'poisoned' air; although the mohair he sorted was very dusty and hairy, dust and hairs were not discovered in the lungs.

This decision followed the presentation of the post-mortem results by one of the three doctors present. The death was determined to have been as a result of blood poisoning.[63] A newspaper article commented on the lack of medical knowledge regarding the causation of the disease, stating that although in this instance the disease appeared to have been caused by blood poisoning as a result of 'miasma', the author did not believe that this was the usual cause of death amongst woolsorters. The focus of the article was on the inhalation of dust and the need for ventilation.[64]

In the preceding weeks, Bell read a paper before the Bradford Medico-Chirurgical Society which was reprinted in the *Bradford Observer* on 16 March 1878.[65] Bell proposed that the wool probably contained 'germs, bacteria, or other living organisms' but that this must be as a result of septic poisons from decaying animal matter, as there was no infectious disease in sheep, goats or alpaca that would cause such a sudden death, though pleuro-pneumonia in cattle was a possibility.[66] Bell wrote that medical men were finding it hard to account for sudden deaths, and so workers were certified as dying from conditions such as 'heart disease, natural causes, perforation of the stomach'.[67] In 1878–9 Bell was influenced by a Leeds physician, John Eddison, to consider whether the disease had any connection to splenic fever in cattle and sheep, a name for systemic anthrax within animals. Eddison had returned from a tour of continental laboratories. He was mentored by Allbutt while he was still working in Leeds and he probably helped to arrange Eddison's visits to French laboratories. Presumably as a result of this conversation with Eddison, Bell's ideas changed radically from those of the previous year. He conducted tests on blood taken from people who had died from the disease. The inoculation of animals with this blood caused the animals to die, and he found anthrax bacilli within their blood. By autumn 1879, Bell was sure that woolsorters' disease was anthrax.[68] In 1880 he forced a coroner's inquest into the death of Samuel Firth on the grounds of death from an industrial accident, for which his employers, Messrs Mitchell and Shepherd, were responsible.[69]

Christine Alvin considers that the cooperation of the medical profession in tackling this disease was 'unique' when compared with actions regarding other diseases in the town at the time, as they were inspired by the 'plight of the woolsorters and their families'.[70] In October 1880, the Bradford Medico-Chirurgical Society established a Commission on Woolsorters' Diseases. The members of the commission consisted of twelve local doctors, including four local medical officers of health. They met thirty times in eighteen months. Twenty-four cases of anthrax in men and animals were examined, and the doctors investigated disinfection, the effects of washing and carried out experiments with fluids and bloods from the living and the dead. Blood from the spleen and fluid from the pleura and pericardium could communicate the disease to rabbits, guinea pigs and mice, and these animals died. Microscopically, the anthrax bacillus was found in malignant pustules and was again fatal to mice.[71] However, there was one medical practitioner in the city who repeatedly threw a spanner in the works, preventing the local medical profession from presenting a unified front in combating anthrax as a bacterial disease. The dissenter was Edward Tibbits, a physician of some importance, with roles as the physician to the Bradford Infirmary and the Bradford Fever Hospital, and previous appointments at University College Hospital and the Brompton Hospital for Consumption and Diseases of the Chest in London.[72]

Tibbits widely published his misgivings regarding bacteriology, locally and nationally. For example, in August 1881, he wrote to *The Times* about 'this so-called woolsorters' disease' to challenge the evidence of the LGB report's conclusion about the links between anthrax and woolsorters' disease.[73] It was not just Tibbits who was initially unconvinced about the link between woolsorters' disease and anthrax. There are indicators of a wider disagreement which began to be resolved by the LGB inquiry of 1881 and the local Commission inquiry of 1882. An article in the *Bradford Observer* in December 1879 complained about the lack of unity within the Bradford medical profession, as some doctors denied that there was a woolsorters' disease at all.[74] However, Bell wrote to the *Bradford Observer* in order to complain about the publication of this article and argued that '[a]ll agree that there is such a disease; that it is a form of blood poisoning; and that it may be easily prevented'.[75] An anonymous 'Medico' (who may of course have been Tibbits) wrote to the *Bradford Daily Telegraph* in May 1880 commenting on Bell's attribution of a death of a woolsorter to 'splenic fever or woolsorters' disease'.[76] This letter was placed in a local paper which was established at a cheaper price than the *Bradford Observer* in 1868, with the aim of being more radical than its rival.[77] 'Medico' remarked:

> As the existence of such a disease is seriously questioned by a very large number of medical practitioners, I would like, with your permission, to ask Dr Bell the question could he detect any difference between what he calls woolsorters' disease and a case of ordinary pneumonia if he were not informed that the patient was a woolsorter.[78]

The 1882 report of the Commission of the Bradford Medico-Chirurgical Society into woolsorters' disease reveals more of the conflict. The report showed that twenty-four cases were studied, which included eighteen in men and six in animals. Five of these human cases were determined to be woolsorters' disease and four were malignant pustule.[79] Experiments were conducted which involved injecting these workers' blood into animals, within which anthrax bacilli were subsequently discovered. The bacilli were also found inside the malignant pustules. The members of the commission came to the following decisions: five concluded that the *Bacillus anthracis* was the cause of both woolsorters' disease and malignant pustule, and the cause of death; five thought that the *Bacillus anthracis* was the cause of the woolsorters' disease and malignant pustule, but were not sure whether it was the cause of death; and one was ill. It is not hard to guess who the one person was who stated that there was a 'Woolsorters' Disease', but

> denied that there was any satisfactory proof that it was internal anthrax or splenic fever, or that the *Bacillus anthracis* had anything to do with it, alleging that that bacterium had no distinctive characteristics as compared with bacteria occurring in other diseases, morbid changes or decompositions.[80]

The minutes of the Bradford Medico-Chirurgical Society reveal Tibbits's continued lack of support for the conventional view of the society regarding germ theory in general, as illustrated in the epigraph for this chapter.[81] Bradford's MOH report for 1884 stated that '[s]everal medical men have evidence, all with the exception of Dr Tibbits expressing the opinion that Woolsorters' Disease is a specific disease due to a distinct bacillus'.[82] In that year, Tibbits questioned germ theory as a whole in his *Medical Fashions in the Nineteenth Century, Including a Sketch of Bacterio-Mania and the Battle of the Bacilli*, where he wrote that germs were 'hypothetical'. He agreed that minute organisms existed, but did not agree that there was any proof that they caused disease.[83] Presumably referring to Pasteur and Koch, but also perhaps to the British bacteriologists and doctors involved with anthrax such as William S. Greenfield of the Brown Institution, Tibbits wrote, 'Some speculative biologists appear to have satisfied themselves on this point, and would endeavour to persuade others that these organisms are dire instruments of destruction to human life'.[84] Indeed, he criticized Pasteur's behaviour at the International Medical Congress in London two years earlier, stating that he acted as if 'he were the only person whose words were entitled to credence' by telling other scientists their ideas were incorrect.[85] Of all theories, germ theory was apparently the most 'fashionable' of the century, and although there may have been some truth in the idea, there was also 'a very considerable amount of error'.[86] In particular, he questioned ideas that anthrax bacilli were the cause of woolsorters' disease.[87] He also critiqued the idea that the concept of the tubercle bacillus meant that tuberculosis was infectious, claiming it was an 'extravagant notion' demonstrated by the apparently rare transmission between husband and wife. He was determined to produce statistics to prove the theory wrong, as he worried that even though the public were surprised by the bacterial method of transmission of tuberculosis, they believed in it.[88] Tibbits also complained about the mistaken identification of pathogenic bacteria which were subsequently discredited as causes of disease.[89]

Conflicts between members of the medical profession were not uncommon in the early years of bacteriology. Many false claims of discoveries of bacteria were made in the 1870s and 1880s, such as Klein's claim to have found the typhoid fever germ; even Koch's anthrax discoveries were sometimes viewed

sceptically in the light of this.[90] Historian Lloyd Stevenson describes many of
the announcements of 'discoveries' of bacteria as 'popcorn'.[91] Indeed, Klebs's and
Eberth's discoveries of the typhoid bacillus barely registered in 1880–1, with a
lack of sufficient evidence to distinguish their claims from those of others. Also,
the bacillus could have just been an additional factor rather than the cause of
the disease.[92] The same kind of issue arose for the tubercle bacillus. Although
the existence of the bacillus was not an issue, its role in causation of the disease
was questioned by a minority of the medical profession.[93] These objections were
similar to the ones which Tibbits suggested: tuberculosis was hereditary, not
easily communicable between people and the animal experiments were not conclusive for transmission amongst people.[94] Indeed, surveys carried out in 1883
by the Cambridge Medical Society and the National Collective Investigations
Committee revealed that most respondents had not seen any signs of communi-
cability.[95] According to Worboys, arguments regarding causality of tuberculosis
were always changing, so bacterial aetiology could have been a passing fashion,
just as Tibbits proposed. In the 1890s these arguments became even more complex when it was discovered that as many as 90 per cent of the population of
some cities may have been infected with tuberculosis, according to the evidence
of diagnostic skin tests and lesions. Indeed, there was 'no closure of medical
opinion before 1900 that consumption was a specific, contagious disease'.[96]

Nevertheless, a lack of consensus amongst the local medical profession in Bradford may have damaged the fight against the disease in the woollen mills. Susan
Jones argues that proof of the bacterial aetiology of the disease was necessary in
order to force employers to change their practices. The lack of unanimity amongst
the Bradford physicians following their research led to doubts cast on the apparently definitive research carried out for the LGB by John Spear and Greenfield.[97]
Along with the national and international context of discoveries of germs being
announced and then proven incorrect, there was also a local problem of a similar
kind in Bradford which did the medical profession no favours when it came to
standing up against local factory owners. Similarly to Klein's announcement that
he had identified a typhoid bacillus and Koch's supposed cure for tuberculosis
(tuberculin), Bell proposed ways of preventing anthrax cases before they were
tried and tested. He argued that the poison could be removed with hot water,
steam or heat, as had been shown by a company who did this and suffered no incidences of disease.[98] In 1879, Bell clearly wrote, in a letter to the *Bradford Observer*,
that anthrax could be prevented in the wool trade. He claimed

> It has been stated over and over again in your columns, that if these noxious wools
> were exposed sufficiently to air or heat, if they were washed or steamed, the risk to the
> sorter, if any, would be slight. The efficacy of these methods of preparation has never
> been questioned ... They are as simple as they are certain to prevent this disease.[99]

He blamed the employers for not carrying out these instructions properly and
for sacrificing human life.[100] Mitchell & Shepherd wrote to the paper to complain about Bell in May 1880, as it had followed his suggestions and yet Bell had

criticized the company for not disinfecting the wool when deaths occurred in its mill. As a result, Mitchell & Shepherd claimed it was going to seek action for libel regarding these comments about its practices.[101] The company's arguments were vindicated in 1881 when the LGB reported that disinfection to prevent anthrax was not possible.[102]

Although the medical profession was generally united in the fight against anthrax, and members were eager to discover the cause of woolsorters' disease and methods of prevention, they also hindered the workers' cause with conflicts and inaccurate claims. The next sections examine how the workers learnt about anthrax, the ways in which they complained, the continuity of ideas formed before discoveries of the bacillus and the means of its transmission, and the way in which workers and doctors united to fight both anthrax and factory owners.

Workers' Knowledge and Actions

There were between 2,000 and 3,000 woolsorters in the Bradford area in the early 1880s, constituting 5 to 10 per cent of employees in worsted factories.[103] Spear, the representative from the LGB, considered that workers closely observed and investigated the disease for many years before it was thoroughly investigated by the medical profession.[104] Knowledge of anthrax as an industrial disease was popularly known in Europe through 'common-sense anthrax epidemiology' since at least 1710, as the cutaneous form was associated with people working with animal products.[105] In 1867, a letter from a former employee at Saltaire Mill called for dissection of deceased workers to reveal the 'secret of their death', because 'men are so impressed by the awfulness of their position'.[106] Indeed, the workers themselves labelled the illness 'woolsorters' disease'.[107]

The scope of the following analysis of the press in order to discover workers' opinions is largely limited to cuttings compiled by Bell and Eurich. However, the scrapbooks are very comprehensive in terms of coverage of the national and local press and include articles and letters regarding various actors and opinions. In addition to the scrapbooks, a few articles have been added from the satirical weekly, the *Yorkshireman*, as these articles did not feature as often in the scrapbooks as those from the *Bradford Observer* and the *Bradford Daily Telegraph*.

Journalists and people who corresponded with the editors encouraged workers to actively complain. In December 1879, a woolsorter wrote to the *Bradford Observer* arguing that the workers should 'agitate' and hold public meetings regarding change in practices to prevent the disease.[108] Although the *Yorkshireman* also suggested that woolsorters should protest in addition to medical men, there was too much risk of losing their jobs; a threat that was 'nearer and more dreaded than the possibility of being poisoned'.[109] The *Yorkshireman* concluded in December 1879 that an enquiry by the LGB was the only sensible action.[110] However, the campaign to involve the workers continued, which suggests that woolsorters were not particularly active in complaining. In May 1880 a correspondent (who Eurich claimed was Bell and, indeed, the letter is signed 'B') wrote to the *Bradford Daily Telegraph*,

> Woolsorters, are you men? Do you consider yourselves of less value than horses? Will you quietly go on being poisoned from day to day, struck down in a few hours with a horrible pestilence, rather than speak out like men, and refuse to touch the abominable stuff till it has been properly cleansed by those whose duty it is to cleanse it?
>
> The cure is in your own hands. Act together, and poor Sam Firth's death may prove a blessing to you all.[111]

From the summer of 1880 the workers' voices were heard as witnesses in the inquests. For example, one woolsorter said he knew of twenty-two people who had died from sorting alpaca or Van mohair.[112] It is hard to find evidence of working-class opinions regarding the disease, making letters to the press a major source. The letters regarding disease causation, listed in Table 3.2, have been discovered in the press. The gap in letters to the press between 1881 and 1897 reflects the gap in deaths from the disease, which is illustrated in Table 3.1. Table 3.2 highlights the letters which were signed by woolsorters. However, the letters from members of the public which display knowledge about bacteria may also have been written by anonymous woolsorters, as some people just signed with their initials.

Table 3.2: Letters written by woolsorters and the public to the Bradford local press regarding woolsorters' disease/anthrax.

Sorter	Date	Newspaper	Idea of aetiology
	25 February 1878	*Bradford Observer*	Blood poisoning
	February 1878	*Bradford Observer*	Dust and 'organic poison'
	1 March 1878	*Bradford Observer*	Smell and dust
✓	2 March 1878	*Bradford Observer*	Dust, mentions stench too, says not especially Van
✓	6 March 1878	*Bradford Observer*	Dust
✓	December 1879	*Bradford Observer*	'Special disease'
✓	December 1879	*Bradford Observer*	Dust
	10 May 1880	*Bradford Daily Telegraph*	Larvae
✓	10 May 1880	*Bradford Observer*	Blood poisoning
✓	3 July 1880	*Yorkshireman*	Smell and dust, but says '[w]hether the disease is germinated by means of invisible animalcula or malaria remains to be proved'
	3 October 1881	*Bradford Observer*	Anthrax bacillus
	13 January 1897	*Bradford Observer*	Anthrax
✓	11 December 1905	*Bradford Observer*	Dust and hair in meals

Source: Photocopy of newspaper cuttings regarding anthrax made by Dr J. H. Bell, until his death in 1906. A few later ones preserved by Dr F. W. Eurich, University of Bradford Special Collections, A15.

The last letter in which miasma is mentioned was sent on 3 July 1880. As Nancy Tomes has shown in her study of knowledge of germ theory in the US, beliefs in noxious air continued despite the media's attempts to dispel such beliefs.[113] Therefore, the cessation in discussions of this concept suggests that the correspondents were quickly acquiring knowledge of bacterial transmission. However, the number of letters is fairly small so perhaps not indicative of the wider workforce. From 1880, two letters refer to the disease being spread by tiny animals, whether 'larvae' or 'invisible animalcula', indicating knowledge of bacterial transmission. The first one, written in May, describes what would have been seen through a microscope. The correspondent wrote regarding the death of Sam Firth:

> Away up at 51, King Street, Bradford, a few days ago ... lay all that remained of Sam Firth ... struck down in four days with a horrible disease, every drop of blood in his body containing millions of living larvae, bred of corruption, inhaled while at work sorting Van mohair.[114]

The second, also written at the cusp of the bacteriological era, is the last letter to discuss bad air, illustrating the overlap in beliefs of disease causation: 'Whether the disease is germinated by means of invisible animalcula or malaria remains to be proved'.[115] In October 1881, the first correspondent mentioned the anthrax bacillus as the cause of the disease.[116] During the next period, in 1905, a woolsorter-correspondent specifically mentioned his concern over meals being contaminated with dust, perhaps revealing an understanding of the intestinal form of the disease.[117] The Bradford workers were warned about this means of transmission; the regulations put forward to the Sanitary Committee as a result of the inquest into the death of a woolsorter in 1880 included a rule that food should not be eaten in the sorting room.[118] However, most of the workers' and families' statements and letters were more concerned with conditions, such as provision of ventilation and fans, and an interest in which type of wool was causing the disease.[119]

Would working-class woolsorters have been able to write these letters themselves? There were opportunities for adult education in the town, including Mechanics Institutes and debating societies.[120] Jones argues that the sorters were 'usually literate and well respected'.[121] Indeed, Spear twice documented that he found these workers to be intelligent. He wrote in his LGB report that 'The woolsorters are a steady, industrious and a most intelligent body of men ... They are skilled with a long apprenticeship and well paid at 31s a week.'[122] In a lecture to the Epidemiological Society of London he also wrote about their intelligence and how they had linked foreign wool with sudden, 'mysterious' deaths:

[B]elieving this, and finding that the fact was not recognized by others, the matter
became one of absorbing interest to them, and they watched the behaviour of the disease – the circumstances and its attacks and its symptoms – with minute attention and
often extraordinary intelligence. The knowledge the sorters themselves thus acquired,
I have found, I need scarcely to say, of the utmost assistance to me in my inquiry.[123]

The press is not the only source of information on worker activism and knowledge of bacteriology. Another example of worker interest in bacteriology is an
entry in the minute book of the Woolsorters' Union in 1901. The union wanted
to ensure that a Home Office promise was carried out, quoting

With every notification by a doctor, or other interested person, of a case of suspected
anthrax poisoning, the medical practitioner attending the patient, shall immediately
forward a specimen of the serum to the bacteriological expert for microscopical
examination, + that in all fatal cases of anthrax a specimen of the serum, taken as
soon after death as possible, shall also be forwarded to such expert.[124]

The union was also involved in the local inquiry involving experimental pathology and anthrax in 1882.[125]

In contrast to workers' knowledge of the disease, a few articles were published
in the press which discuss their ignorance of transmission of the disease. In 1884
it was pointed out by the *Bradford Observer* that workers needed to take care of
their health in the workplace. A woman had had a small wound through which
the 'poison' had entered. The journalist argued that the workers needed to be
instructed on how to look after themselves, particularly regarding the problems
of working when they suffered from external wounds.[126] In an inquest in June
1888, it was revealed that a man who died had slept on bags of wool and had felt
'cold and shivered' when he awoke. Apparently, this was a 'piece of recklessness
or ignorance which [was] almost inconceivable after the warnings that have been
given'.[127] Indeed, Carter notes that in Kidderminster, workers were often blamed
for cases of anthrax with accusations that they lacked 'self care' and did not see a
doctor soon enough.[128] So the press argued that workers still needed to be taught
how to be careful in the possible presence of anthrax.

'Here He Lies, Killed by Van'[129]

Examining letters regarding Van mohair reveals the way in which workers discussed the cause of the disease and how it could be prevented. Workers were
particularly concerned about the smell and dust which emanated from the
foreign wool. For example, one letter complained: 'The smell of it alone is "horrible, horrible, most horrible." One bale of it will pollute the atmosphere of a
large room.'[130] A letter to the *Bradford Observer* written on 1 March 1878 perhaps stimulated campaigns against this material, complaining about dust, and

encouraged workers to refuse to work with this wool. 'Yorkshire' cited workers' comments as his source:

> Mohair is, generally speaking, harmless; but the Van mohair, to use a sorter's phrase, omits on opening 'a stench like a grave.' The sorter having got over the smell, breathes an atmosphere so dense from dust and infinitesimally small hairs that his neighbour at the next board cannot see him; his nostrils are stuffed, his lungs are clogged ... [a few days later ... [I]n terrible agony the poor fellow has met the 'Woolsorters' death.' Is there any necessity for this terrible sacrifice of human life? It is distinctly traceable to this particular Van mohair. The sorters pronounce it damaged, and say that for this reason it is cheaply bought up in those days of bad trade, and can be worked up into very nice-looking material, which can be sold at a good profit. But what a profit – the life of the poor woolsorter![131]

The writer proposed improvements in mills such as fans, ventilation and washing wool before it was sorted, and suggested that the workers should refuse to work with Van mohair if employers did not comply. Two quick and critical responses from woolsorters followed in the newspaper, one written the day after the letter from 'Yorkshire' was published. One suggested it was up to the medical profession to determine the cause, and the other stated that although '[the Van mohair's] stench and dust is almost unbearable', he knew of sorters who had died never having come into contact with this type of wool.[132]

An inquest in early March must have encouraged these ideas about Van mohair, as there was much discussion about the dangers of the wool, even though the death in question occurred five weeks after sorting Van, followed by the sorting of alpaca.[133] An article in the *Bradford Daily Telegraph* on 5 March 1878 cannot have helped the reputation of Van either, commenting that it was no surprise that the wool was infected due to recent reports about the 'habits' of inhabitants of the area from which it came, and that there were rumours of the 'prevalence of malignant epidemics' amongst those people.[134] A letter which followed this publicity about Van was written to the *Bradford Observer* on 6 March, personifying the wool, with the signature 'Van Mohair'. The letter complained that sorters had worked with Van mohair for years with no ill consequence and, as with a previously mentioned letter, argued that there were sorters who had died who had not been in contact with Van. 'Van Mohair' demanded statistics to show whether mohair sorters were more at risk and if sorters were in more danger than any other employees.[135] Mitchell Brothers were only too happy to provide statistics (the letter signed 'Van Mohair' could of course have been written by the company), which showed that the death rate had not increased since Van mohair was introduced.[136] However, the belief in the harmful nature of Van continued, and the paper read by Bell before the Bradford Medico-Chirurgical Society, which was printed in the *Bradford Observer*, repeated the idea again, and added camel hair to the danger list.[137] Letters and arguments from sorters concerning Van also continued to be published in the press.[138] Bell surveyed the workers of Bradford regarding what they thought about 'the belief that Van Mohair + Alpaca are more dangerous than others'. They responded that the dry dust and the smell were the cause.[139] In 1905 Legge reported that Persian wool was at least as likely as mohair and Van mohair to contain the anthrax bacillus.[140] Nevertheless, as discussed earlier, Van mohair was perceived as particularly dangerous by the sorters and inspired the Harden Mills strike.

A common complaint about Van mohair was the dust which was dispersed from the bales. Table 3.2 reveals that workers focused on dust as a cause both before and after woolsorters' disease was linked with bacteria. Bell argued in his 1878 paper that dust was not the cause of woolsorters' disease as occupational illnesses which introduced dust to the lungs tended to be chronic rather than acute and dust-reducing fans did not reduce the incidence of the disease.[141] Yet dust was naturally associated with microbes. Workers' claims that dust was the cause may not have merely resulted from concepts of the disease which predated bacteriology, involving the 'mechanical' rather than infectious harm caused by dust.[142] In 1908, Eurich re-emphasized the role of dust as a result of brittle dried blood on the fleeces.[143] The occupational health specialist, Thomas Oliver, argued in 1916 that dust was dangerous in general, because microorganisms may be attached. In the wider field of occupational health, early in the twentieth century confirmation also came that diseases like silicosis resulted from inhalation of fine silica dust.[144] However, Stark has argued that a sanitarian approach to dust in the sorting rooms illustrates the continuity between ideas before and after the introduction of bacteriology in the knowledge of woolsorters' disease. One of the measures which stemmed from the questioning of witnesses at the Firth inquest was the control of dust, along with methods aimed at achieving disinfection, including lime and carbolic acid. Spear's recommendations also included collecting and burning dust. Despite these measures, Stark asserts that as bacteriological knowledge of anthrax increased, methods of controlling anthrax focused on the factory environment rather than the battle against bacteria. The Bradford Rules which were displayed in every sorting room did not even mention bacteria.[145]

Working Together: Doctors, Sorters and the Press

Anthrax in Bradford received heightened levels of publicity because of collaboration between the medical profession, the woolsorters and the press. Employees' knowledge of woolsorters' disease was well regarded by doctors, who encouraged workers to complain. Bell thoroughly interviewed the workmen and his notebook reveals the format of his questionnaire. Questions included: 'When [bale] opened any traces of vermin, animal parasity?', 'Have new comers been affected more readily than others?', 'Are the men physically equal to other classes?', 'Have you sometimes a stinking bale wh: had not been damaged by water?'[146] This relationship was symbiotic as the workers hoped that doctors would join their campaign. For example, one of the sorters' letters to the press, written at the end of 1879, suggested that the doctors should collaborate with them, despite their lack of consensus on the topic. He proposed a public meeting where 'the practical woolsorter and the scientific medical men would be brought together, and publicity would be obtained'.[147] The sorter also wanted an LGB inquiry. However, other letters from workers show more subservience towards doctors. One of the first letters from sorters suggested that the whole matter of the cause of woolsorters' disease should be left to the medical profession to decide.[148] Another said the sorters should 'look to' men like Bell for help.[149]

These were not the only exchanges that occurred. As Susan Jones and Philip Teigen have proposed, the woolsorters of Bradford were perhaps more likely to see doctors and alert them to woolsorters' disease than those working elsewhere as a result of the Woolsorters' Society, which was established in 1818. Along with 'controlling employment and policing workers' in order to ensure good working relationships with the factory owners, the Society provided sickness benefits if the sorter had been examined by a physician or had a certificate from a steward. Without this the worker would also be fined by the society for missing work, providing extra incentive to visit a doctor.[150]

The local press can also be seen to have worked with the majority of the doctors and the workers in fighting anthrax in the wool trade. The press printed details of the inquests, including how lawyers and manufacturers argued about charges against them, as well as disseminating bacteriological research. Additionally, the press published both sides of the arguments between manufacturers and employees and among conflicting doctors.

The woolsorters communicated with the *Bradford Observer* more than the other local newspapers, even though the *Bradford Daily Telegraph* aimed to be more radical. The *Bradford Observer* quickly debated and published news of bacteria and used this in a campaign supporting the workers, which berated the proprietors of the woollen mills. The minutes of the Medico-Chirurgical Society reveal that readers were assimilating this knowledge; it was noted on 19 September 1883 that doctors needed to keep up their understanding due to lay knowledge.[151]

On 3 March 1878 an article was published which claimed to summarize the state of knowledge about the disease at this time from the evidence given at an inquest, speculating that mechanical irritation by dust was the cause, and arguing against the idea that animal matter or 'poison' could result in disease:

> It is difficult to see how the bad influence from decayed animal matter in packed mohair can penetrate through the lungs and poison the blood more readily than the poison from many other substances regarding whose deadly effects we hear nothing.[152]

Two days later, the *Bradford Daily Telegraph* stated that the disease was caused by 'infected' wool, 'somehow in a pestiferous condition'.[153] These articles perhaps provoked contributions from the MOH, Harris Butterfield, whose letter was published three days later. He wrote to the *Bradford Observer* stating that the disease was caused by an 'animal poison', and later in the same month the article was published which documented Bell's paper on Woolsorters' Disease at the Bradford Medico-Chirurgical Society.[154] The *Bradford Observer* then considered the disease was caused by inhaled germs or bacteria, but the actual cause had not yet been identified.[155] In December 1879, the *Yorkshireman* told its readers about the 'myriads of germlife disclosed by the microscope' in Van mohair and in May 1880, the *Bradford Observer* announced:

> We believe that, so far as scientific investigation has gone, the form of blood-poisoning known as woolsorters' disease is caused by the presence in the decaying matter attached to the wool, of germs the fatal vitality of which can only be destroyed by methods similar to those used in fever-hospitals for the purification of infected clothing.[156]

The day before the newspaper had warned readers about beliefs in miasma:

> Woolsorters' disease is not due simply to a bad smell; it comes from living parasitical germs, bred of corruption; and until these have been killed by a high temperature the danger remains in the wool. The peculiar sickening odour is merely a warning of danger, and is not itself the danger.[157]

On 1 October 1881, a correspondent suggested reprinting an article from *Nineteenth Century*.[158] He told the workers about much of the information contained in any case, writing 'Sir, – In the *Nineteenth Century* for the current month there is an article by Dr Carpenter on "Disease Germs" which is worthy the earnest perusal of every thinking man, and especially of the woolsorters of this district'. The correspondent summarized the article, including information such as the mediums in which germs grow, the investigations carried out by Pasteur and others, ways of preventing diseases in the future and work on vaccination of animals. Perhaps both he and the author of the article were over-zealous, as he wrote, associating the vaccine with that for smallpox, 'This discovery, to use the doctor's own words, "must to the woolsorters of Bradford prove a most important boon, if they can be led to understand the value"'.[159] The article was apparently 'teeming with information of vital importance to them ... which may possibly lead to a more simple and far more efficacious method of extirpating the "woolsorters' disease" than that at present exists'.[160]

The transition from miasma and dust to bacteria in journalists' discussions of the cause of woolsorters' disease was very quick. Even Pasteur said in 1881, that it was in that year that the theory was generally accepted, presumably referring only to the medical profession, as he connected this comment with the International

Medical Congress in London. Codell Carter has questioned why this year was the tipping point and speculates that it may have been as a result of the impact of Pasteur's anthrax inoculations.[161] Indeed, in terms of the British general public, the events at Pouilly-le-Fort were reported in *The Times* by their correspondent in France, who was invited to the farm to witness the experiment. The experiment was reported as being very important for agriculture.[162] This would have given publicity for bacteriology in general, especially with regard to anthrax.

Anthrax was not the only pathogen which exposed Bradford to highly publicized ideas regarding laboratory medicine. In 1886 Bradford was struck by another disease which crosses the animal/human boundary, and for which there was a new innovation through laboratory medicine. Five months after Pasteur had first tried his vaccine treatment for hydrophobia, eight dog-bite victims from Bradford set off for Paris where Pasteur was offering free treatment for hydrophobia for anyone across the world, attracting hundreds of patients within weeks. The MOH at the time was Thomas Whiteside Hime, an enthusiast of laboratory science. He appealed to Pasteur for vaccines but instead was asked to send the patients to Paris. On 13 March, seven patients, including five children, went to Paris with Hime. They were later followed by another man. These people had been treated locally by cauterization but after the death of another victim on 11 March, the dog was proven to be rabid and a subscription was started to raise funds for the dog-bite victims' travel. The Bradford eight were not the first from Britain to visit Paris for the treatment; five people went between November and January, but the press did not pay much attention. Neil Pemberton and Michael Worboys argue that the Bradford group captured the media's attention after the reporting of four boys from New Jersey in the American and British press in December.[163] News of the Bradford group reached the national press, with *The Times* reporting on 13 March that two patients were on their way to Paris to see Pasteur and that there was fundraising in place to try to enable the others to join them.[164] Pemberton and Worboys discuss the Bradford 'brigade's' transformation into national celebrities through the local and national press.[165] Like anthrax, although hydrophobia was a frightening disease, in comparison to more common infectious diseases it had a much lower death rate amongst the British population. However, the third highest number on record was sixty deaths in 1885.[166] New developments in microbiology and relatively high incidence of the disease kept laboratory medicine in the minds of the citizens of the town. Indeed, five more local residents visited Paris for treatment in the spring and summer of 1886, including Hime who wounded himself while experimenting in the laboratory.[167]

To summarize, the cause of woolsorters' disease was established by collaborations between local doctors and workers. They used the press in order to disseminate these communities' ideas and to encourage each other to complain to employers about the risk of the disease, but local newspapers were also active in distributing information about the disease. However, the factory owners also used newspapers in order to complain about their representation in the press and about the inconsistencies in Bell's arguments. Yet, even though there were inquests, and voluntary rules were introduced in order to encourage factory owners to care for their workers, there were no official channels in order to complain about the disease or to gain compensation, a topic which is discussed in the last section of this chapter.

Compensation Claims and Lawyers' Use of Bacteriological Knowledge

The second period of anthrax cases and deaths during the late nineteenth century began in 1897 and was accompanied by arguments for compensation. Anthrax was included as a notifiable disease in the Factory and Workshops Act of 1895. It was listed as a dangerous disease, included as a form of poisoning alongside chemical substances such as phosphorus, arsenic and mercury.[168] The Workmen's Compensation Act of 1897 interested lawyers in the disease as it led to arguments as to whether infection by the anthrax bacilli could be regarded as 'accidental' in the terminology of the act. Therefore, when anthrax returned to Bradford at the same time as this act, there were new agendas and conflicts.

In 1903, a death from anthrax in Kidderminster resulted in a local case for which compensation was awarded in 1905. Brinton's, the manufacturer, appealed against this decision, but it was upheld in the Court of Appeal by Lord Halsbury's argument that if an employee was cut by an object, resulting in tetanus, this would be seen as an accident.[169] Carter presents the conundrum of why a Kidderminster case was the first to result in compensation and not a Bradford case, considering political activism and the involvement of the medical profession in Bradford. There were only five deaths from anthrax in Kidderminster between 1900 and 1914.[170] There had also been campaigns for compensation in Bradford and in 1905 several cases were awaiting the result of the Kidderminster case. Two of them were specifically mentioned in the press as awaiting the outcome. As a consequence of the Kidderminster case, compensation was awarded for four cases in Bradford between April and June.[171]

In the Kidderminster case, the wording of the Workmen's Compensation Act of 1897 was challenged; on the one hand anthrax was seen as an 'incident' of the job, and on the other as an 'accident', a definition which was required for compensation by the 1897 Act. The House of Lords ruled in favour of the latter claim because an accident had been defined in a previous case as 'used in the popular and ordinary sense of the word as denoting an unlooked-for mishap or an untoward event which is not expected or designed'.[172] As Julia Moses has argued, the Latin root of the word 'accident' is 'something that happens' and in Italian compensation legislation it was further defined as having a 'violent cause', whereas in British legislation it was simply presumed to be a 'transparent' term. Despite the more flexible British terminology, compensation for bacterial disease occurred in Italy first. Italian cases included compensation for bubonic plague amongst dockworkers in 1904.[173]

The 1897 Act allowed compensation to be paid for accidents in certain workplaces with a 'no fault' principle, in order to provide an inexpensive and sure way of gaining compensation in comparison with the Employers' Liability Act of 1880. The first recorded case of a successful claim for compensation in the workplace was for injury to a worker as a result of negligence and was brought to the High Court in 1837. However, there may have been out of court agreements before this. The 1880 Act did not provide protection for accidents, and no financial help was provided for costly cases, and so before 1897 workers and their families could not sue employers for damages caused by contracting anthrax.[174] This lack of infrastructure for complaints did not stop discussion of compensation in the 1880s in Bradford.

Claims of negligence began in Bradford in 1880 when Bell blamed the employers for woolsorters' disease in a death certificate. An article in the *Bradford Daily Telegraph* related how the employer was responsible to the state for accidents as a result of negligence. The journalist questioned what was to be done.[175] Compensation was not discussed as a serious option for sorters' families in the late nineteenth century, although this situation was lampooned in the *Yorkshireman* in September 1880:

> What will people want compensation for next? Is it not enough that railway companies must pay for passengers smashed for lack of continuous brakes, that employers must pay for workpeople smashed by rotten boilers? No, it is not, for Mr. Walter Dunlop has sent in a claim to the Bingley Local Board for £50, the value of three cows, which died of Woolsorters' disease! It seemed that the cows had grazed in a certain field, and that the field was watered by certain sewage, and that the sewage contained certain washings of certain woolsorters'-disease wool. Now, as woolsorters don't get compensation for dying in a much less roundabout way than the cows, the Board didn't see Mr. Dunlop's claim.[176]

Rather than compensation for families if the breadwinner died, woolsorters were paid high wages. As previously mentioned, Spear commented on the high wages paid to sorters. However, the *Yorkshireman* reported in June 1880 that this increased wage was 'by no means sufficient' to support a family if the main breadwinner died.[177] In July, a sorter complained in the *Yorkshireman* about the low wages woolsorters received as a result of a surplus in people wanting jobs, but that people working with Van mohair sometimes chose this job as they were paid more money.[178] Indeed, in an inquest in 1880, one worker stated that sorters of

Van mohair were paid more than those sorting English wool, even though it did not require more skill. He believed that it was because the wool was unhealthy.[179]

The letters which encouraged workers to complain and organize themselves indicate that they were not instinctively doing so. Stark has argued that the Bradford Trades and Labour Council, representing employees, encouraged woolsorters to seek compensation under the Employers' Liability Act, providing financial support and organizing demonstrations.[180] Otherwise, inspiration seems to have come from national policy.

At the very end of the nineteenth century, attempted claims for compensation accompanied the resurgence in the disease. The claims followed the inclusion of anthrax as a notifiable disease in the 1895 Factory and Workshops Act and the new 1897 Workmen's Compensation Act. Bartrip suggests that the Workmen's Compensation Act of 1897 was a result of the Conservatives competing for working-class votes. However, the Act did not necessarily result in the improved safety demanded by the trade unions but apparently led to companies insuring themselves for accidents instead.[181]

In 1899, a workman tried to claim compensation for half wages for the time when he was away from work to have a lump cut out of his neck following a pimple becoming infected with the anthrax bacillus. Compensation was not awarded as the judge agreed with the respondent's lawyer that it could not be proven where the workman came in contact with the bacilli.[182] A case reported in July 1900 in the *Bradford Observer* was of importance as it attempted to establish anthrax as an accident under the terms of the Workmen's Compensation Act, just like the subsequent Kidderminster case: '[t]he whole question in the case was what is an accident? There was no definition in the Act.' The argument was that there was a difference between an accident and a disease. The prosecutor's lawyer said that an accident was something not designed nor caused by negligence, and that if accidents caused by fright and gas had been included, why not disease. The judge retorted, 'If the germ had flown out and cut his lip I should agree that it was an accident. One might say in all cases that disease was accidental in a certain sense.'[183] In his summing up of the case and decision in favour of the defendant, the judge further explained why the disease could not legally be considered an accident, using his interpretation of a disease caused by a pathogen:

> To bring an injury within the meaning of the Act there must in his opinion be an accident which caused immediate injury to some part of the body. A bacillus settling on a cut or being breathed into the lungs did not itself do any injury, it was only when it grew and multiplied that the diseased condition of the body was caused, and he did not think in the first instance till that disease ensued, there could be said to be any injury.

He quoted another case – the words of Chief Justice Cockburn in the case of *Sinclair* v. *The Maritime Passengers' Assurance Company* – where disease was apparently caused by climate. It was decided that some violence or casualty should be involved in order for the definition of an 'accident' to be applied. Cockburn had remarked, 'It is true that exposed to the same malaria one man escapes and another succumbs yet diseases thus arising have always been considered not as accidental, but as proceeding from natural causes'. Continuing his argument, Judge Bompass commented that

> In the present case it has been contended that either the fact of infected fleeces being sent over might be considered as an accident or that the germs flying from the fleece, lighting on a scratch or wound on the deceased and giving him the disease, were like a spark flying in the eye of a blacksmith, and are an accident. I am of opinion, however, that there is no personal injury by accident within the meaning of the Act. I think that the Act intended to distinguish between disease and accident, and that infection from germs was not intended to be treated as equivalent to an external injury, such as the injury from a spark. Moreover, I think the sending over infected fleeces with others was not an accident, but the ordinary course of the trade. Any other interpretation of the Act would bring most diseases caught by workmen during their employment, such as scarlet fever caught from a fellow-workman or inflammation of the lungs caught from some defect in the heating apparatus, within this statute, and this, I think, the provisions of the Act as a whole show cannot have been intended.[184]

Unfortunately for workers and their families, defending lawyers used knowledge of bacteria and medical debates at the turn of the century in order to argue that not everyone was affected by bacteria and that bacteria and disease were natural occurrences. This is contrary to the later 1904 Italian case where a sudden attack of bubonic plague was classed under a definition of violence and to the successful argument which won the 1905 Kidderminster case when anthrax was compared to tetanus.

Conflicts and Agendas – Using Knowledge of Bacteriology in Late Nineteenth-Century Bradford

There were complex interrelations and conflicts regarding industrial anthrax in Bradford. Some actors encouraged each other, some complained about each other, or even publicly disagreed with those within their professional group in the case of Tibbits and the medical profession. The relatively low number of deaths from anthrax compared to the public outcry suggests that the disease was used in worker/employer conflicts in the Bradford area, influenced by socialism and by reformist newspapers. The purpose of studying Bradford at the beginning of the bacteriological age is to understand the extent of dissemination and public engagement with bacteriology at this time. Rather than a revolution with the introduction of bacteriology, there was a mixture of ideas with a concentration on smells as well as new ideas of bacteria, just as Tomes has seen when examining concepts of disease in everyday life in the USA.[185] This is not surprising as the workers were aware of sensing strong smells rather than invisible bacteria. Yet knowledge about bacteria still appears to have spread very quickly in Bradford.

The timing of the early discovery of the anthrax bacillus is highly significant for the early reception of bacteriology in Bradford by the medical profession, the press and workers. Anthrax research was crucial for confirmation that specific bacteria caused specific diseases, with Koch and Pasteur's discoveries and their spat occurring at the same time as the campaign to prevent anthrax in Bradford. Koch's discovery was influential, as the size of the bacterium, its life cycle and its reproductive powers matched the sudden appearance and the symptoms of anthrax.[186] Indeed, it was a conversation about a colleague's tour of continental medical laboratories that gave Bell the idea of anthrax being the cause of woolsorters' disease. Koch's discovery was acknowledged by many, as the large bacteria were analogous to parasites such as worms which were already accepted causes of disease. Also, Koch created a 'virtual theatre of proof' as his demonstrations were supported by drawings and microscopy. Worboys describes his discovery of anthrax as a pathogen as an 'event' unlike many of the other true or false claims of bacteria discoveries over the following years.[187]

Additionally, socialism within the town probably led to early attitudes of blame and responsibility for disease, as did the Conservative Party which vied for working-class votes through the Workmen's Compensation Act. However, except for the strike at Harden Mill, workers were not very active in complaining about disease in the workplace. This perhaps demonstrates a general expectation by workers and families that death in the workplace was not unusual in the late nineteenth century. This attitude changed in the very late nineteenth and early twentieth centuries. This also shows that, as the *Yorkshireman* pointed out, the risk from anthrax was not as threatening as the risk of losing a job as a result of complaining.

In examining the creation of the Bradford and national rules for regulation of the wool factory environment, Stark has argued that 'esoteric bacteriological debates' were not utilized.[188] Yet, for other purposes, this chapter has shown that at least some skilled manual workers were very capable of understanding bacteriology, that doctors valued and respected the knowledge of the workers regarding disease, and that lawyers could use their knowledge of medical science in order to defend employers. With so much additional publicity about hydrophobia, the people of Bradford were unusual in being exposed to knowledge about innovations in laboratory medicine at such an early stage.

The next chapter explores a much less well-organized workforce in east London, before examining the general risk of catching anthrax within the capital. As anthrax in London has barely been discussed by historians before, the focus is much more on regulation and attempts to protect workers than has been the case in this chapter. Whereas the plight of woolsorters in Bradford was well understood by 1905, and exemplified by a number of local and national strategies, from the local Bradford Rules which included ventilation and sanitation in the factories in 1884, to the Anthrax Prevention Act and disinfection station at Liverpool in 1919, the fate of people working with leather was often neglected. However, within the period studied in the current chapter, the Bradford Rules did not necessarily cause abatement of the disease, as despite a decline in reported deaths during the late 1880s and early 1890s, anthrax returned in the last years of the nineteenth century. Indeed, the highest number of fatalities occurred in 1906 and nationally, the number of deaths in the wool trade was highest in 1916 and 1917.[189] The subsequent, but questionable, success story following the establishment of the Anthrax Investigation Board for Bradford and District in 1905, the Home Office Departmental Committee on the prevention of anthrax in 1913, and the eventual discovery of a working method of disinfection by G. Elmhirst Duckering, has been acknowledged in detail by several historians and there is no need to retell the whole story here.[190] From 1913, Duckering, His Majesty's Factory Inspector, carried out experiments in order to find a way of disinfecting wool. After over a hundred trials he reported a working twelve-stage process which included the use of 2.5 per cent formalin, a substance which had been suggested by Eurich in 1906.[191] This led to the establishment of a disinfection station in Liverpool, funded by tariffs on the imported fleeces, but it immediately ran into significant debt. In addition, its effect was limited as only selected fleeces were disinfected.[192] Chapter 4 explores why a focus on east London is essential for a full understanding of anthrax in Britain, and in particular for the aims of this book. Yet, the danger of bacteria was not always well publicized and acknowledged in London, even when an international organization became involved.

4 ANTHRAX IN LONDON: LEATHER, ZOO KEEPING AND SHAVING BRUSHES, 1882–1932

Work in a tannery is among the most nauseating of trades. To be constantly engaged in handling the hides and skins of animals that may have been dead for months is anything but a picnic.[1]

A 'popular penny-weekly' quoted in *Leather World*, 1912

Between January and April 1882, three men died at the London tannery, Messrs Barrow Brothers. They were all infected with anthrax by poor quality hides which had arrived from Shanghai. Additionally, a man who transported the hides to the tannery suffered from a mild case of the disease. More bundles from the same batch were sent to Paris, as Barrow Brothers no longer wished to work with them. In Paris, seven men were struck by 'blood-poisoning', two of them fatally. Then, in July, there were three more cases of the disease at Butler's Wharf, London, among men who were working with more of Barrow Brothers' rejected hides.[2] Clearly, outbreaks of anthrax in the London leather trade could be just as serious as those in Bradford. Of the cases of anthrax in the leather trade between 1904 and 1909, forty-three were in London. The next highest number was in Liverpool where there were twenty-one, yet the mortality rate was much lower in London.[3] Were employers, employees or doctors responsible for the much lower case mortality rate in London? This chapter argues that historians of anthrax and labour need to pay more attention to the risk of the disease from leather and the role of Britain in the international debates regarding this problem. London is the focal point for new historical research into the workplace experience of anthrax, and innovative bacteriological experiments and observations took place in the capital city, beyond those which were commissioned by the LGB to investigate anthrax in Yorkshire.

In his Milroy Lectures for the Royal College of Physicians in 1905, Thomas Legge announced that between 1899 and 1904, 261 cases of anthrax had been reported to the Home Office: eighty-eight from wool, seventy from horsehair, eighty-six from hides and seventeen more from other trades such as work with horns and rags.[4] Statistics from the Chief Inspector of Factories show a total of 943 cases of anthrax within the wool trade, 600 within the hides and skins trade, and 235 cases among those working with horsehair between 1900 and 1939.[5] Although there were 50 per cent more cases within the wool trade, the significance of the disease within the hides and skins trade should not be dismissed. Indeed, between 1903 and 1909, case mortality from anthrax in the leather trade surpassed that in the wool trade.[6] Following the opening of the Liverpool disinfection station, the number of cases among those working with wool or hides and skins was very similar: 259 cases in the wool trade and 239 cases among those working with hides and skins between 1922 and 1939.[7] The priority given to studying the dangers of pulmonary anthrax in the wool trade has resulted in historians neglecting serious study of the incidence of cutaneous anthrax among those working with hides and skins: a risk which was very similar to that amongst wool workers, except for a few years in which the incidence of the disease in the wool trade was particularly high (1905, 1911 and 1916–18).[8]

Anthrax cases in the hides and skins trade almost exclusively began with a malignant pustule. This particularly surprised one of the bacteriologists who investigated anthrax in the leather industry, as despite there being a lot of dust in the tanneries, especially when dried hides were handled and manipulated, there were only two cases of the pulmonary form between 1873 and 1894 and only two recorded cases of intestinal anthrax during the period 1903–9.[9]

The external form of the disease could be as frightening as the internal form, yet the pustule clearly gave a warning and treatment was sometimes successful, meaning that outcomes were much better. One man's experience illustrates how an apparently simple pimple could quickly escalate into systemic infection and death. In 1884, Isaac R., a 48-year-old warehouseman, had been working in the leather works and handling hides from Bombay. One of his 'mates' asked him, 'What's the matter with your neck? It looks red.' On the next day a distinct swelling had developed, but he carried on working until six in the evening. After finishing work he had difficulty in breathing and he vomited two or three times. He arrived at Guy's Hospital in a cab on the next day but died a day later, within four days of his friend noticing the red mark.[10] Although the mortality rate was not as high, the cutaneous form of the disease could still be swift and deadly.

The preoccupation with woolsorters' disease may result from the use of this term in considering late nineteenth- and early twentieth-century anthrax. According to Stark's research on Bradford, the term 'woolsorter's disease' became associated with the human form of anthrax, whereas 'anthrax' was linked with the contamination of bales of wool. By the early twentieth century, 'woolsorter's disease' could also be used to describe the external form of the disease as well as the pulmonary condition.[11] Even a local London newspaper reported the Southwark coroner Dr F. J. Waldo as opening an anthrax inquest by explaining, 'anthrax was a wool-sorter's disease, derived from foreign wool, that coming from Persia, China, and the East', before introducing the case of John Callow, a waterside labourer who had been involved in unloading hides.[12] Stark also notes the conflation of the term 'woolsorters' disease' with 'anthrax' by historians.[13] This amalgamation may have led to the lack of attention to anthrax amongst leather workers within the historiography of anthrax.

Most historians of anthrax have momentarily mentioned the risk for those working with hides but only a few have paused for thought.[14] The most extensive comments on hides and skins in London, running to at most two pages, are used to show that Bradford was not the only location in which anthrax was an industrial disease, rather than studying the importance of the disease in the capital in its own right. In particular, Bartrip and Stark focus on Spear's sequel to his LGB report for Bradford. In 1882, Spear investigated cases in wharves, warehouses and tanneries, linking the cases to hides from China and other ports and suggested that caustic lime could be used to disinfect them.[15] Even though Bartrip perceives that the incidence of anthrax in the capital actually led to the Home Office committee which investigated anthrax in the workplace in 1895, the focus of the committee remained on wool in Lancashire and Yorkshire. However, the special rules regarding hygiene and ventilation in workplace environments, in addition to facilities for changing clothes and dining, which were issued in 1897, were for employees involved with both woolsorting and hide and skin.[16]

Another reason why the London story has been skimmed over is that biographies of the disease tend to focus on the story of the Anthrax Investigation Board and the subsequent committee which was appointed to discover and enact a means of disinfection of wool. In particular, the focus on wool is perhaps driven by the triumphant success with disinfection and the resulting Liverpool disinfecting station. In contrast, today there is still no solution to the disinfection of leather in the early stages of the tanning process without causing irreparable damage. After the opening of the station in 1921 and the debatable success in reducing cases within the wool trade because only limited types of wools were disinfected, the historiography of anthrax generally leaps forwards to biowarfare during the Second World War.[17] This results in neglect of the story of international collaboration which tried in vain to find a method for disinfecting hides and skins.

Internationally, the contamination of hides and skins with anthrax was much more significant than infected wool. Table 4.1 illustrates that in France and Germany, hides and skins were much more likely to cause disease. Although the statistics were not comprehensive for the United States, between 1910 and 1917, there were 222 deaths from anthrax, 42 resulting from skins, 14 from hair, 5 from wool, 28 among transport workers and 42 in agriculture. Therefore, except for Britain, the major economies of the time experienced a much more significant problem with anthrax within the leather industry than in woollen mills.[18]

Table 4.1: Statistics of human cases of anthrax submitted to the International Labour Office by the French, German and British governments, 1910–20.

	1910	1911	1912	1913	1914	1915	1916	1917	1918	1919	1920	Total
France												
All industrial cases	54	42	38	66			157			48		405
Hides and skins	35	25	20	30			91			33		234
Hair	5	2	5	3			7			5		27
Wool	10	8	6	26			32			6		108
Germany												
Hides and skins	110	89	96	81	74	19	9	6	18	6		508
Hair	15	19	22	18	16	6		1		3		100
Wool			1		1		1					3
Great Britain												
All industrial cases	51	64	47	70	55	49	106	99	68	57	48	714
Hides and skins	14	20	8	19	13	18	18	29	14	16	17	188
Hair	6	8	7	5	5	2	6	1	4	3	5	52
Wool	28	35	31	43	29	26	80	65	49	34	25	445

Source: International Anthrax Commission, *Memorandum Circulated By the British Representative* (London: HMSO, 1922), International Labour Organization Archive, Hy 501/1/2.

Melling has ventured into analysis of the interwar period, along with co-authors Mortimer and Carter. The focus of these articles is on the wool trade, only briefly acknowledging hides and skins.[19] Mortimer and Melling account for the decline in anthrax in the early 1930s, in particular in hides, as corresponding with the economic depression and reduction in trade rather than medical improvements.[20] Bartrip briefly explores the interwar period, continuing Mortimer and Melling's work to dispel the myth that anthrax virtually disappeared in Britain following the opening of the Liverpool disinfection station. There were 2,145 cases of cutaneous anthrax between 1900 and 1956, 261 being fatal. A total of 987 of these cases occurred after 1921.[21] This suggests that incidences of anthrax were little more than halved by the disinfection station. One of the reasons why anthrax fatalities still occurred was that it was difficult to eliminate the spores from dry hides.

This chapter questions why the people unpacking and working with foreign hides and skins did not gain the same attention as people working with wool, despite a few contemporary reports, and why this has resulted in a neglect of the topic in the historical account of anthrax in Britain. Risks to leather workers are contextualized within wider threats of anthrax in London workplaces and homes, with the purpose of expanding current studies of the bacterial disease beyond the focus on wool and biowarfare. Hence, the chapter examines a variety of risks in the capital, from working with horses to working in the laboratory, before focusing on an unusual outbreak which struck both elephants and humans at London Zoo. The chapter connects to the next part of the book, which examines wider communities, by investigating the history of knowledge of bacteria derived from publicity about the contamination of shaving brushes during and after the First World War. Although the workers' voice is often absent regarding anthrax in London, this chapter endeavours to explore their understandings of disease which are relayed in publications by doctors, and to examine dissemination of bacterial concepts in the press. It also seeks to show incidences where pioneering bacteriological research was attempted in relation to anthrax within the leather industry and zoo keeping, as well as the network of bacteriologists who were involved in controlling anthrax carried by shaving brushes.

Manufacturing Leather: A Risky Business

In 1883, the *British Medical Journal* published an article, 'Anthrax in Bermondsey'. The article set the scene by discussing the study of anthrax in Bradford but stated that over twenty cases of anthrax had recently been seen at Guy's Hospital: 'The frequency with which this disease has been met with at Guy's Hospital is, no doubt, due to its position in the neighbourhood of the great tanyards, leather manufacturers, warehouses, and wharves of Bermondsey and Rotherhithe.' The article argued that the LGB needed to think about the subject which was causing some 'consternation' among workmen engaged in the skin warehouses.[22]

The LGB had already noticed the incidences of anthrax in east London. In 1882, Mr Payne, the coroner for Southwark, asked the LGB to investigate the cases of malignant pustule in hide warehouses and tanyards in Bermondsey where London's Hide and Skin Market was based. Spear investigated incidences of the disease, and this report was a sequel to his LGB report on anthrax in Bradford in 1880. When Spear questioned the people engaged in the trade, he found that they believed that the disease was caused by mineral poisons such as arsenic and corrosive sublimate which were used in the course of curing hides. Even the coroner's juries came to this conclusion. Meanwhile, the clinicians and pathologists at Guy's Hospital proved that the malignant pustules resulted from an infection by anthrax.[23] Spear's report alone illustrates the importance of bacteriological tests at Guy's in changing the approach to this industrial disease.

The term 'hide' refers to the skin of larger animals, including cows, camels and horses. 'Skin' alludes to smaller animals including sheep, goats, rodents and carnivores. Spear believed that anthrax was much more likely to be transmitted by hides as wool and hair were usually removed from sheep and goats, reducing the value of skins and the likelihood of their exportation to London.[24] Animals were skinned in the place where they died or were killed. If a hide was to be immediately tanned no curing was necessary, but if there was a delay, for example due to the hide being transported, the moisture had to be removed or antiseptics applied in order to prevent putrefaction. The hides imported from China and most of those from the East Indies and Africa were dry and those from Australia, South America and European countries were usually wet. Skins were usually dry regardless of their origin. The dry hides and skins from China, East India and the Cape were considered to be the most likely to be contaminated. Hides and skins were usually cured with antiseptic dips or by rubbing in dry antiseptic powders such as napthalene. 'Dry-salted' skins could be treated by being coated in saline earth, which may have included sodium sulphate which had antiseptic properties. Sodium arsenite was also used in order to prevent putrefaction and deter insects. However, these chemicals did not necessarily prevent anthrax. Anthrax was seldom conveyed by the wet hides as anthrax bacilli were dispersed in dust and because the wet hides present adverse conditions, even for the spore form of the bacilli. Although wet hides were safer, they were heavier and therefore more expensive to transport. Nevertheless, from 1898 to 1909 between 46 and 69 per cent of imported hides were wet.[25]

At the quay, hides and skins were either unloaded or transferred to the wharves via barges. They were then wheeled to the dock sheds or warehouses. The bales were opened and examined and brokers took samples which they inspected on behalf of the tanners wishing to purchase them at auction. These warehouses were dusty environments. After the bales were purchased, hides were transported on the railways to their destination.[26]

As with the wool trade, the earlier stages in the sorting and processing of hides and skins were the most dangerous. Within the tannery, most cases of anthrax occurred among the people who soaked dry skins and hides so that they became flexible and would absorb tannin, or among those who cleaned the hides (which might have been contaminated with animal dung or blood). The former task included the use of antiseptics such as carbolic or boric acid, or sodium sulphide. This process may have been an effective way of disinfecting leather before it reached employees who worked in later stages of the manufacturing process, as the anthrax spore could have returned to bacillus form in these wet conditions and therefore have been killed by the disinfectants.[27]

The other early manufacturing process, using a solution of lime to dissolve the hairs, involved scraping off the hairs with a knife. The lime also served to plump up the hides. The hides were soaked in lime for several days, being momentarily removed each day so that the lime deposits could be 'plunged'. Sometimes the hides were transferred to a new pit of lime if the solution became too foul. Despite the belief that up to a week in milk of lime would disinfect the hides, bacteriologist Constant Ponder conducted tests which showed that this was not the case, confirming earlier research in Britain and on the continent. As the disease was very rare beyond this part of the process, Ponder believed that the scraping process must have also contributed to the removal of the spores. Transmission of anthrax was not impossible beyond this stage and spores had been found later in the tanning process; there were cases among shoemakers, and horses had apparently contracted the disease from their leather reins. Ponder argued that in these rare cases there was a possibility that the disease was caught from elsewhere or that the finished leather had come into contact with unfinished hides. The liquid effluent created in the tanning process also caused trouble as it could pass into streams from which cattle might have drunk, a problem which had occurred in Market Harborough in Leicestershire in 1904.[28]

Situated close to the River Thames in east London, Guy's Hospital gained an increasing reputation as the destination for London anthrax sufferers. Within the *British Medical Journal* (*BMJ*) in June 1884, John Poland, the surgical registrar at Guy's Hospital, wrote that there had been thirty-six cases of anthrax admitted to the hospital, including twenty within the last two years. Eight of these people died, but twenty-eight recovered. The malignant pustule was excised on admission. However, the *BMJ* reported that it was too early to claim that excision was the reason for people's survival.[29] A subsequent report highlights the concentration of cases at Guy's: between 1873 and 1894, 85 of the 119 cases diagnosed in London were treated there.[30]

In 1890, N. Davies-Colley published an article in the *Guy's Hospital Report* which documented the thirteen cases he had treated since 1883 and the remedies with which he had experimented. He emphasized that these were only a quarter of the cases which had been treated in the hospital as a whole. Davies-Colley had been excising the pustules and experimenting with various ointments for the wounds, such as zinc chloride, before he read an article about the use of ipecacuanha by Edwin Muskett, a doctor working in South Africa. The thrust of Davies-Colley's article was that he not only decided to apply this externally, but to prescribe it internally so that it would be slowly absorbed in the alimentary canal. Ipecacuanha is a plant originating from Brazil which was already used to treat dysentery. Like Muskett, Davies-Colley had a very good success rate with the patients he treated with this botanical drug. All five patients had recovered, even one patient who was in grave danger, having had significant problems with breathing. In examining the action of the drug, the hospital bacteriologist found that it destroyed pure cultures of the anthrax bacilli but not the spores. This was not a problem inside the body as the bacteria would not be in spore form. This led J. W. Washbourn and Percy Evans from the Bacteriological Laboratory to conduct more bacteriological research experimenting with substances which might make good disinfectants or successful antiseptic creams, preventing the bacilli and spores from reproducing. They tested substances which Koch had been working with, judging that the substance he advocated, corrosive sublimate (which was another name for bichloride of mercury), was not safe for the skin in concentrations which were sufficient to destroy the bacilli. Each of the creams which were tested had a disadvantage of some kind, for example creolin was effective but was too opaque so that the wound could not be observed.[31] Therefore systematic bacteriological research into therapies, disinfectants and antiseptics was occurring in London in addition to Bradford.

The anthrax patient's voice is difficult for the historian to discover in London but Davies-Colley's report provides an insight into their fears, as illustrated already by the fate of Isaac the warehouseman. Spear's primary recommendation in his 1882 report was 'the dissemination of a knowledge of the danger amongst workmen, and of the precautions that are necessary to be observed in the first symptoms of the attack' because current remedies were much more likely to work if the sufferer reported to a hospital quickly. Spear praised Dyster, Nalder and Company, one of the largest hide brokers, for issuing instructions in plain, simple language to all of their employees. However, this simple language meant that the words 'bacilli' and 'bacteria' were replaced with 'poison' resulting from a diseased animal, so the terminology and concepts of bacteriology were not disseminated.[32] The poster designed for the woollen mills in Bradford also excluded reference to organisms.[33] Information on the disease had not been rapidly disseminated to all of the tanners in the mid-1880s. Walter S. worked at Messrs Beresford and Company in St Olave's Wharf. In April 1884, he noticed a pimple and he scratched it even though he had been handling hides from China and Bombay. It swelled rapidly. He told his doctor that he had 'never heard of any similar affection among the workmen'. In the same year, George R., an employee at a skin brokers, waited six days before going to Guy's, firstly applying a sticking plaster, then waiting as one of the pimples rapidly swelled for two days, before feeling faint and suffering from stomach pain. A salter in the hide trade working at Culverwell, Brook and Cotton, tried to 'get rid of' his pimple with two ounces of Epsom salts. He waited six days before appearing at the hospital in June 1888. Although William J., a labourer, reported to Guy's a day after he noticed a swelling, this was because he was suffering from anthrax for a second time in March 1889. Yet some workers were more aware; Michael W. reported a day after he had noticed a swelling in March 1884.[34]

Between 1883 and 1893, there were fifty cases of anthrax at Guy's Hospital, with an average of one death per year over the latter seven years. This inspired the MOH for London County Council to launch an inquiry.[35] William Hamer, the Assistant Medical Officer, produced the report on the 119 cases which had occurred in the capital over the previous twenty-one years, including 26 cases which were fatal. As the number of cases increased, local doctors' knowledge had developed. Despite the declining imports of dry hides, Hamer agreed with Spear that improving workmen's knowledge was the best way of defending themselves against the disease.[36] The *BMJ* argued that compulsory notices regarding the risk of anthrax were the best way to enforce this and that this knowledge had already been dispersed.[37] Inquests continued to be reported in the *BMJ* and in 1896 there was a notable outbreak in Bermondsey amongst sorters of foreign hides.

Between 31 January and 4 February, six cases were admitted from one factory. The patients were treated with excision and ipecacuanha, and all cases were 'progressing favourably'.[38]

From the beginning of the twentieth century, some deaths from anthrax were documented in the national press.[39] A key article showed the problem of casual labour and the confirmation of transmission of the disease. Frederick Beeby, a waterside labourer in Southwark, died from anthrax in 1913. At the inquest, Mr Ringrose, HM Inspector of Factories, said it had been impossible to trace his last place of employment even though he usually worked at Red Lion Wharf. However, the representative for one of his last workplaces gave evidence at the inquiry and argued that wharves were not factories and were therefore exempt from the laws in place for anthrax. However, he had still displayed the required notices for factories and provided washing facilities. The evidence did not prove where the disease was contracted. Beeby had cut himself shaving and then a pimple had formed. The serum was administered to him at Guy's but the disease was too advanced.[40] If anthrax could not definitely be assigned to a particular workplace it is very doubtful that a worker would be entitled to Workmen's Compensation, which was paid by the employer who may or may not have been insured for claims.

The coroner at most of the cases in Southwark, Dr F. J. Waldo, called for more public education about anthrax as the number of cases was increasing even though the Home Office had instilled regulations. He also argued that the regulations for factories and workshops should be extended to wharves. At one inquest in 1914, the jury returned a verdict of accidental death, exonerating Messrs Major and Field but arguing that the Home Office needed to extend the laws to wharves within which animal products were received.[41] Again, in 1920, the jury added a rider at another of Waldo's inquests that regulations needed to be extended to wharves. In that case, a warehouseman who worked with Russian skins had died but the warehouses were not included in factory regulations.[42] Despite these manufacturers being exonerated from blame at inquests, other tannery owners noted the harm which anthrax was causing to employees. The proprietor-editors of the journal *Leather*, subsequently *Leather World*, appear to have been very concerned about the danger and tried to attract wider attention to the problem.

Between 1911 and 1913, *Leather World* published news or feature articles on anthrax every month or two. In 1913, the weekly trade journal protested, 'the recurrence of fatal cases certainly suggests the time has come for the Government to take further steps in the matter, as it is a disgrace to our civilization that human life should be sacrificed in this way'.[43] Feature articles highlighted problems with the inspection of dangerous hides in China and the recognition of dangers from hides originating in Africa.[44] Discussions of anthrax had become so common in the journal that perhaps it resulted in the use of a bacteriological metaphor when discussing another topic, with the 'bacilli of unrest' used in reference to strikes about wages.[45]

The epigraph for this chapter is extracted from an article which critiqued the portrayal of the leather trade in a popular penny weekly. The editors had received several letters from those engaged in the leather trade; although some people had found the article amusing, other correspondents were worried about people being dissuaded from working in the industry. The *Leather World* front-page piece argued that there was no need to worry, since people were recruited from the 'hardiest section of the community, the mere look of a tannery or currying establishment being enough to frighten the more squeamish lads'. However, the penny-weekly feature had been alarmist and *Leather World* protested about the tone of the report. It had been published under the heading 'Dangerous Trades: Slaves of the Tannery', and described leather work as 'among the most nauseating of trades'. Although this article went on to complain about problems such as the chemicals which resulted in hands and arms being 'pickled almost as thoroughly as the dead skins themselves', anthrax was made most prominent as the discussion began by highlighting the risks of that disease:

> Unfortunately, there exist a good many people who are quite without conscience where money is concerned. Such folk do not hesitate to skin a beast that has died of anthrax, in spite of the terrible risk to themselves and others.

Mention was made of the fact that there had been fifty-four cases in the trade during the previous year and that eleven had been fatal. Although the author of the article for *Leather World* conceded that anthrax was a problem which 'naturally, we all deplore', he wished that the small proportion of cases could have been compared with the large number of people involved in the trade and that it had been noted that medical discoveries were reducing the number of deaths when sufferers were promptly treated.[46]

Another recurring topic was the need for a means of sterilization of hides and skins. The trade was currently 'drifting along in an aimless manner' with regard to the problem of disinfection.[47] A. Seymour-Jones's research into disinfection appeared several times, along with the argument that the government needed to know about his study of the value of bichloride of mercury.[48] In 1911, in particular, the trade journal published a few articles which educated readers about bacteriology and disinfection. James Scott wrote a two-part feature about 'anthrax germs' for the 'layman, devoid of highly scientific terms' because of the 'general interest taken in the important matter of Anthrax and its prevention'. He illustrated his article with two images – the magnified bacilli in a petri dish and the structure of skin. Scott explained how the anthrax bacilli usually succumbed to other bacilli in the process of putrefaction of hides and discussed research into disinfection.[49] Another two-part article by J. H. Yocum discussed experiments in disinfection, and disseminated the research by bacteriologists including Robert

Koch.[50] Disinfection was also discussed in the light of the dangers from Chinese hides in addition to the dangerous effluent from tanneries.[51]

Leather World was obviously aimed at the leather trade but the periodical was subtitled 'the weekly journal for all engaged in the Leather, Tanning, Hides, Skin, Wool and Allied Industries'. News from Bradford was also included within its pages. Otherwise, there is so little other content regarding the wool trade that these articles on anthrax appear to be for the sake of comparison with the leather trade. News about decisions regarding the responsibility for individual deaths in Bradford was published, along with updates about developments in research into disinfection of wool.[52] In addition, the periodical followed news of anthrax more widely, publishing stories about anthrax as a result of other trades, in various places.[53] Other articles provided news of deaths derived from reports by the MOHs in London or from inquests.[54] They reveal the manner in which it was thought that the workers had contracted the disease. For example, a dock labourer in Poplar had scratched himself with his fingernail.[55] Further afield, not only labourers were affected but also a manager of a leatherworks in Eastwood, West Yorkshire, who had only been in the building with Singapore hides and had not actually handled them.[56]

Although *Leather World* did much to publicize and deplore the problems which were resulting in anthrax in the leather trade, the journal occasionally blamed the workers. In the case of the manager who died in Eastwood, the Home Office notice about the safe handling of hides had not been displayed, even though the Home Office had been very active in sending these out to tanneries meaning they were a 'perfectly familiar object'. However, the author of 'The Deadly Anthrax' argued that the men who worked with hides and skins were

> callous, and neglect even the most rudimental precautions for their safety – even down to personal cleanliness ... It seems almost an impossibility to expect workers to take measures for their own protection – even if they are to some extent effective ... do not look for an early improvement in their ways.[57]

Indeed, when the wife of a labourer at St Olave's Wharf, Bermondsey, died, the onus was placed on her husband as she was put in danger because he brought the bacilli home on his clothes. The inquest discussed the long overalls which workers refused to wear and the problem that they did not wash their hands before they went home.[58]

The Home Office Departmental Committee, which was appointed to examine woolsorting, proposed in a report in 1897 that hides and skins also needed special rules. In 1899 rules were drawn up and served on seventy-six firms. There were objections, especially as wet hides were included, but also because the workers were expected to wear high-necked and long-sleeved overgarments which fitted closely at the wrists. The rules were soon amended to refer only to work with dry skins and hides. In 1901, the Home Office held a conference and decided that special rules were needed for hides arriving from China or the west coast of India. These regulations included washing facilities, provision of sticking-plasters and dressings for small wounds, keeping food and clothing away from the area, and reporting a small wound or scratch to the foreman and discontinuing work until it was properly dressed. The rule about the high-necked garment had also been removed on the proviso that the bales were not to be carried on the shoulders. The revised rules also included instructions that workers were to be constantly reminded of the dangers of anthrax.[59]

In contrast to the blame placed on workers, *Leather World* praised the St Olave's Wharf labourer for his intelligence in spotting his wife's disease. The man had sent his wife to hospital, suspecting anthrax. At first she was turned away as presumably she had not explained her husband's concerns. Her husband sent her back to the hospital and she was admitted once the disease was recognized. She was prescribed the serum and the pustule was removed. However, the patient died.[60]

Meanwhile, the problem had also come to the attention of the Worshipful Company of Leathersellers and, in 1911, bacteriologist Constant Ponder, MD, wrote a report for the London guild. This report was publicized in *Leather* in December 1911.[61] Ponder's report announced that case mortality was higher within the hide and skin trades than in the wool, hair or bristle trades. Ponder argued that the active measures in Bradford had led to this statistic, particularly as a specially appointed doctor was inspecting any suspected incidence of the disease as soon as possible, with a further expert carrying out bacteriological tests. This resulted in treatment for the disease at the earliest possible stage, and therefore lower case mortality.[62]

As noted earlier, the case mortality rate in London between 1904 and 1909 was lower than Liverpool: 18.4 per cent in comparison to 42.5 per cent. As several cases occurred in Bermondsey each year, Ponder believed that the men were better acquainted with the risk and the early appearances of the disease. They knew when they should seek medical help at Guy's Hospital where surgeons were familiar with the disease. By 1911, it was a rule at the hospital that when any patient presented with a 'boil', he was asked if he had had any contact with hides and skins and an immediate bacteriological examination was made. If the examination revealed anthrax bacilli, he was immediately treated. If the test was negative, the patient was kept under close observation and had to regularly report until the boil had healed. Observations were accompanied by further bacteriological examinations. Thus, Ponder argued, case mortality in London was lower than in Liverpool.[63] In addition, the ethos of first-aid training had been gradually introduced into shipping between the 1880s and 1908. After thirty years of campaigning, compulsory training was introduced so that there would be ambulance-trained officers on vessels travelling abroad. This influx of newly trained men, some of whom had undertaken voluntary training in the years before, might have spread knowledge among the men who were interchangeably employed on the ships and in the docks.[64]

Dockworkers were particularly at risk from the disease, as illustrated by Table 4.2. A total of 59 per cent of cases occurred in wharves, docks and warehouses. Ponder had not expected this as he thought that the highest proportion of cases would be within the sorting process. Therefore, disinfection needed to occur before the hides and skins even reached the tanneries.[65] This risk for dockworkers might have been underreported. The problem with the Factory and Workshop Act of 1901, was that medical practitioners did not have to notify cases in other workplaces.[66] Risks were not restricted to people who worked directly with leather; in Liverpool a dock labourer had died even though he had not handled any hides and skins, illustrating the dangers of cross-infection in shipping and the docks.[67] Ponder speculated that there might have been a high incidence amongst dockers because they carried the bundles near their necks and because their work resulted in more skin abrasions. When Ponder investigated this idea, he discovered that the carrying of bales on labourers' shoulders was not common practice. In any case, there were more pustules on the left-hand side of faces, and on the occasions when bales were lifted near the neck, they were usually placed upon the right shoulder. Out of 113 cases, 93 began with a pustule on the head or neck. There was a good chance that this was related to shaving and acne leaving open wounds for the bacilli to enter. When there were longer gaps between contact with anthrax and ill health, this could be explained by spores remaining under the fingernails until the workman scratched an open wound.[68] *Leather World* reported that one method of protecting dockworkers was to provide them with gloves which could be disinfected every day in a solution which was used to try to kill the anthrax spores.[69]

Table 4.2: Cases and case mortality occurring at different stages
in the handling and manipulation of hides and skins.

Where contracted	Cases	Cases fatal	Percentage fatal
Wharf and dockyard	48	13	27.1
Warehouse	12	1	8.3
Tannery	42	10	24.4
Contracted indirectly	11	5	45.5
Totals	113	29	25.6

Source: C. Ponder, *A Report to the Worshipful Company of Leathersellers on the Incidence of Anthrax amongst those Engaged in the Hide, Skin and Leather Industries, with an Inquiry into Certain Measures Aiming at its Prevention* (London: Worshipful Company of Leathersellers, 1911), table 11, p. 27.

Within the rest of his report, in addition to researching anthrax in animals and humans, Ponder presented the findings of his experiments with proposed methods of disinfection, particularly those suggested by the International Commission for the Preservation, Cure, and Disinfection of Hides and Skins.[70] A. Seymour-Jones from England was the chair of this commission. As noted earlier, he had been experimenting with soaking dry skins in a solution of bichloride of mercury and formic acid; then the skins were soaked for an hour in a solution of salt water.[71] Ponder thought that Seymour-Jones's method had the most promise. Formic acid sped up the soaking of the hides and bichloride of mercury had a germicidal role. This resulted in a quick transformation of dry hides into wet, and the addition of a germicide was useful as trials revealed that most samples of leather were disinfected although a few bacilli from the sample were still able to kill guinea pigs.[72] In the USA, a certificate of origin or disinfection was required for imported leather, although it was not usually possible to prove where exactly the hides had come from. Hence, bichloride of mercury was supposed to be used for disinfection, but this damaged the hides, suggesting that Seymour-Jones's methods would not be useful. Disinfection of skins was not straightforward as '[s]uccessful tanning depend[ed] upon delicate and complicated chemical and biological processes'.[73] A whole range of other possible methods of disinfection were tested by a variety of researchers, including solutions, powders, ozone, light and electricity. These had varying effects, for example blue and violet light actually encouraged the reproduction of bacilli and the formation of spores, although intense gaslight retarded the growth of the bacilli.[74]

In the absence of a successful means of disinfection, Ponder was keen to find ways of preventing the disease. Although gloves were worn in tanneries, there was a danger that the moisture which would gather in the gloves could develop spores into bacilli. Therefore, the gloves needed to be boiled or disinfected after every use. Ponder also thought that fingernails should be kept short to prevent scratching.[75] Legge suggested that the workers should be provided with warm water for washing as the use of cold water might result in chapped hands. They also needed to wash their hands in flowing water so that the water for washing did not become contaminated.[76] Yet, the risk from anthrax bacilli was even higher among those people who handled the leather bales before they reached the tanneries. Legge argued that for practical reasons, disinfection could not occur before a bale was opened in the warehouse, in comparison to wool and hair for which experimental methods of disinfection were discussed.[77] The MOH for the Port of Liverpool, Dr Hanna, suggested that the bales needed to be covered in canvas and lifted mechanically.[78]

At the time of Ponder's report, it would be three years until the discovery of a successful method of disinfecting wool. However, in 1913, the wool trade succeeded in gaining significant attention and funding through the Home Office committee and the commitment to find a means of disinfection. The leather trade did not receive the same attention. In 1911, Ponder had tried to show the seriousness of the problem in the hide and skin trade by comparing the value of imports to cases across the affected trades. His statistics revealed that the same value of goods resulted in one case of anthrax in the wool trade and two cases in the hide and skin trade. Despite the lowest number of cases being within the hair and bristle trade, there were ten cases for the same value of raw material as in the wool trade. In order to reduce the variables, Ponder decided to examine the cases of anthrax amongst those involved in tanning or hide and skin workshops as he thought that dockers were involved in other jobs and so the time which they spent with leather was difficult to calculate. He actually found that manipulating the data in this way meant that the risks were lower for tanners than for the wool or horsehair workers because of the large number of people involved in the leather trade.[79]

Ponder's report and his experimentation with the statistics he collated serves to highlight that the risk for dock labourers was significant. Table 4.3 illustrates how dangerous it was for dockers to handle hides and skins, in comparison to other materials which could be infected with anthrax. The number of cases which occurred among dock labourers may have resulted in less attention to the disease, because they were less likely to submit compensation claims. It is probable that the number of cases was actually higher as notification of the disease was not compulsory outside the factories and workshops. Since 1906, an employee who suffered from anthrax had the right to compensation if he suffered from disease or injury and disablement reducing his ability to earn full wages or if he was suspended from employment on account of a disease, or his family had a right to claim if he died.[80] The casual employment of dock labourers probably reduced their likelihood of claiming and their lack of unionization in comparison to Bradford may have meant that they were also unaware of their rights to compensation. Imagining the life of a docker in late nineteenth- and early twentieth-century London helps us to understand why this may have been the case.

Table 4.3: Cases of anthrax and deaths from the disease reported among British transport and warehouse workers, 1900–31.

Kind of material	Dock and wharf labourers	Warehousemen (dockers)	Warehousemen (factory or other premises)	Car, cart and motor drivers	Railway porters	Total
Wool	5 (2)	25 (7)	31 (8)	2	1	64 (17)
Horsehair and bristles	1	2 (1)	5 (2)	3 (1)	-	11 (4)
Hides and skins	148 (37)	33 (1)	21 (1)	8 (1)	1	211 (40)
Bones	4 (1)	–	–	–	–	4 (1)
Grain	2	–	–	–	–	2
Not ascertained	15 (5)	2	3 (1)	2 (1)	1	23 (7)
Totals	175 (45)	62 (9)	15 (3)	15 (3)	3	315 (69)

Note: Number of fatal cases in brackets.

Source: J. C. Bridge, Factory Department, Home Office, London, to Dr Carozzi, International Labour Office, Geneva, 25 January 1932, International Labour Organization Archive, Hy550/4/0.

A dockers' day would begin by jostling for the employers' attention at the dock gate or a hiring point. The social commentator Henry Mayhew wrote about the thousands of men at the docks, jumping on the backs of others so that they could be seen, perhaps shouting their own names to the employer to try to be recognized.[81] This was a degrading way to find employment every single day. As a result of the clamour for work, a couple of the London wharves were known for their violence. If a docker found work, an employer paid him only for hours worked or by tons moved; men could be employed by the day or by the half-day to suit demand. Employers had no responsibility for their welfare.[82] This system of employment explains the experience of Frederick Beeby, the waterside labourer at Red Lion Wharf who was mentioned earlier; it had been impossible to determine where he had been working before he became unwell.

The desperation of the docker for work, and therefore the lack of power for negotiation, is explained by the figures of unemployment in this sector of the economy. The 1908 Royal Commission of the Poor Law reported that 'pauperism was three times higher amongst dock labourers than the national average.'[83] Until 1912, anyone could try to find work in the docks, so the role attracted the unemployed from other trades too, a problem that was addressed after the First World War in order to prevent demobilized soldiers taking dockers' jobs. Unemployment was estimated to be at 30 to 40 per cent before 1912 and 26.5 and 39.8 per cent between the world wars.[84] The surplus of men meant that if some went on strike there were always others willing to take their place.[85]

Victorian reformers believed that the dockers were demoralized, leading to 'drinking, brawling, gambling and a happy tolerance of each other's vices'. Rather than seeing this as a 'counter-culture', they identified it with the dockers being 'loafers' with no intention of full employment at the docks. In stark contrast to Spear and Bell's descriptions of the woolsorters of Bradford, Mayhew wrote, 'This class of labour is as unskilled as the power of the hurricane. Mere muscle is all that is needed, hence every human locomotive is capable of working there.'[86] With such job insecurity and the lack of respect for their work, dockers perhaps did not perceive anthrax as their primary concern.

If dockers specialized in particular goods and tasks, this improved their chances of employment. It took years to learn the 'stevedores' art of stowing goods', how to carry sacks on swaying gangways, or to grade commodities like tea.[87] Despite the anonymity experienced by Beeby, some dockers built up relationships with employers who recognized their particular skills. Those at risk from leather would probably have worked in quay porterage. These men tended to specialize in particular cargo which they not only unloaded, but which they sorted, counted, weighed, marked and packed onto carts, trucks or barges.[88] Ponder does not appear to have realized that the dockers were probably involved in sorting goods, not just carrying them, which seriously exposed them to the risks of anthrax.[89]

The limited trade unionism in the docks developed amongst skilled labourers first, with strikes occurring in 1872, 1889 and 1911. The 1889 strike resulted in men being employed by the half-day rather than per hour, meaning they only had to attend two calls for work per day rather than waiting for work all day.[90] The First World War further improved their conditions, as the value of the dockers' labour was finally valued as of national importance. Dockers were exempted from military service and their labour was supplemented by 15,000 troops. Unions were allowed representation on national and local Port and Transit committees from 1916. Some men were employed on a permanent basis. In 1889, only 19 per cent of dockers had this kind of contract, however in 1894 a total of 30 per cent and by 1904 a total of 41.2 per cent were employed with permanent contracts. However, the men who had permanent contracts tended not to get involved in union politics as they did not want to risk losing their jobs with so many other men clamouring for work.[91] Therefore, with such insecurity for permanently employed men in addition to those employed casually, recognition of the value of their labour and better contracts were the focus of union leaders rather than the prevention of anthrax.

These difficult conditions are reflected in compensation statistics. Casual dockworkers were entitled to Workmen's Compensation but they were only awarded payouts which reflected their income, so if they only managed to find employment two or three days per week, their claim was in proportion.[92] Despite the relatively high risk of anthrax among dockworkers who handled hides and skins, compared to the rest of the leather trade, the number of Workmen's Compensation claims was not high (see Figure 4.1). This is especially the case for the period on which Ponder focused, in which he argued that case mortality was highest among those working with leather. In 1911 there were thirty-five cases of anthrax in wool and twenty-nine in other trades, whereas there were twenty compensation claims regarding people working with wool and only ten for people working in other trades – this is an approximate comparison of cases and claims as they might not have occurred within the same calendar year. The relatively high number of claims from those working with wool in the first ten years after anthrax was added to the Workmen's Compensation Act of 1906 reveals that there was much more knowledge about compensation in the wool trade than in the hide and skin trade and at the docks. By the late 1920s, claims made in relation to anthrax by people working in trades other than the wool industry were catching up. Yet, even when cases in other trades surpassed wool, the number of claims did not. The numbers of cases reported to the factory inspector in 1928 were sixteen for wool and twenty-four for other trades, and yet there were still sixteen claims for wool and only eighteen for other trades.[93]

In 1932, W. J. O'Donovan published an article in the *BMJ* commenting on the power of the surgeon's certificates for dermatitis and how they constituted a 'legal fact'. Even if O'Donovan's diagnosis was suspected to be incorrect, the recording of his opinion on a certificate was binding against expert dermatologists in court. In 1919, twenty-seven cases of anthrax were certified. O'Donovan argued that the Workmen's Compensation Act had led to a 'compensation complex' for sufferers of skin diseases.[94] There was certainly a high proportion of claimants from the woollen industry but some people who were afflicted in the other trades did not bother to claim. It may have been that they only suffered from minor cases which were successfully treated but considering the attention to the early treatment of anthrax cases in Bradford, this should certainly have been the case in that city too.

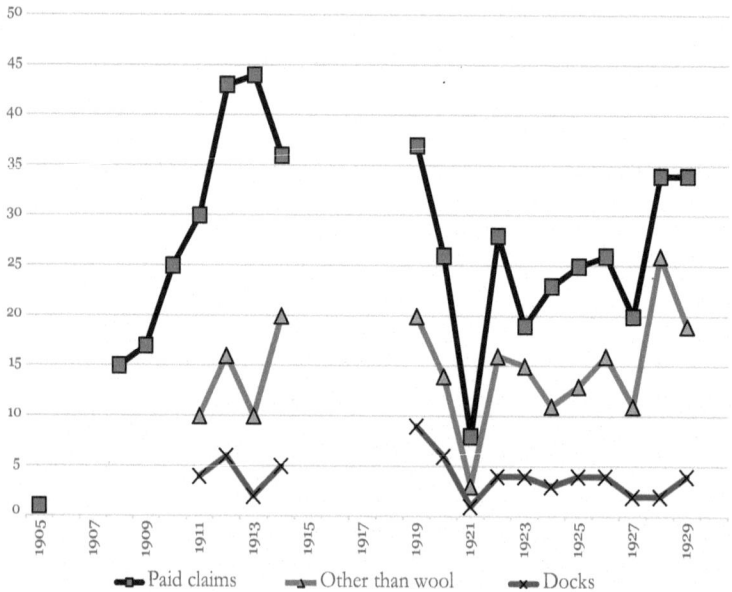

Figure 4.1: Number of compensation claims per year for anthrax cases, specifying claims for people working in all trades, in trades other than wool, and in the docks. Source: Home Office, *Statistics of Proceedings under the Workmen's Compensation Acts, 1897 and 1900, and 1906, and the Employers' Liability Act, 1880*, House of Commons Sessional Papers, 1905–29 (London: HMSO, 1906–30).

Anthrax in London and the International Labour Organization

The British government's attitude towards international bacteriological research into anthrax in the leather industry highlights the state's emphasis on the risk of the disease from wool rather than leather. The first international committee to study anthrax in a global context was established by the International Society of Labour Legislation in 1912. However, it failed to discover a method of disinfection before Eurich and Duckering devised a means for wool in 1914.[95] The Anthrax Prevention Act of 1919 empowered the British government to prevent importation of goods which were suspected of being infected. This was followed by the establishment of the wool disinfection station in 1921, which resulted in Britain leading the world in the fight against industrial anthrax. However, the leather industry attracted international attention.

Founded in the aftermath of the First World War, the International Labour Organization (ILO) was situated in Geneva. In 1920, anthrax research and regulation was one of six recommendations at the International Labour Conference in Washington, DC.[96] The prevention of the disease in the workforce was also one of the important questions at the ILO meeting in Geneva in 1921.[97] A resolution was adopted at this third ILO conference to investigate disinfection of wool, hair, hides and skins, as well as the challenging topic of prevention of anthrax amongst animals. The Americans were noted to have a particular interest in prevention of anthrax in hides and skins.[98]

The topic of anthrax was addressed by an Advisory Committee of the ILO which met in London in 1922. The committee members came from Australia, Belgium, France, Germany, Great Britain, India, Italy, Japan, South Africa, Spain, Sweden and the USA. Despite the USA abstention from the League of Nations and this associated organization, a representative was sent.[99] At the outset of the 1922 meetings, Albert Thomas, the French director of the ILO, expected that there would be an international resolution regarding disinfection of wool. He predicted that proposals regarding hides and skins would be presented for the 1924 Conference Agenda, and that the prevention of anthrax in animals would probably have to be addressed by the League of Nations Health Organization or the International Institute of Agriculture in Rome.[100]

The topic of leather was the focus of the eighth sitting of the committee. The committee agreed that the danger of infection from hides and skins, was 'from the international point of view, as important as, if not more important than, the danger from wool and hair and the text [of the recommendations] should include some finding to that effect'.[101] After ten sittings in London and a visit to the disinfection station in Liverpool, the committee concluded that the focus had mainly been on recommendations of disinfection of wool and hair, with resolutions regarding research into disinfection of hides and skins, horns, hoofs and bones, as well as prevention amongst animals.[102] Yet one of the conclusions of the report was that '[t]he possibilities of spread of infection from infected hides and skins are even greater than in the case of wool, and the need for dealing with the danger under some international agreement is at least as strong'.[103] Thus, the topic of anthrax and leather became the first research agenda of the Industrial Hygiene Committee, one of the research divisions of the ILO.

In 1923, the committee came to a decision that the research into hides and skins should be undertaken in collaboration with the Health Committee of the League of Nations. The Health Committee appointed a 'Mixed Committee' of six members, three from the Health Committee and three from the Correspondence Committee on Industrial Hygiene. The goal was to test the methods proposed by 'various experts', to see if they worked in industry, and 'to study their results from the technical, economic and bacteriological point of view'. The Industrial Hygiene representatives were from Britain, France and Italy. Legge was invited to represent Britain but declined due to pressure of work, recommending Edgar Collis, who was professor of preventive medicine at the Welsh National School of Medicine in Cardiff.[104]

Although the British were looked to for leadership of this committee, there was considerable reluctance. Malcolm Delevigne, who was responsible for factories at the Home Office and was the British representative to the ILO, expressed surprise that Britain was being asked to take a role in the expert committee:

> We are of course interested in the question of disinfection of hides and skins but I was not aware that it was largely at our instigation that the Labour Office pressed for the appointment of the Mixed Committee! Would that the Office had displayed as much zeal about the disinfection of wool![105]

Delevigne probably would not have been as candid in expressing his view had it not been that the deputy director of the ILO, H. H. Butler, was British. Butler responded that the Committee on Anthrax had made the decision to conduct research, but that it was the British government who initiated the ILO anthrax investigation in 1922.[106] This was duplicitous, as in July 1923 Butler had written to Luigi Carozzi, who headed the Industrial Hygiene Committee, to inform him that Delevigne was keen on the hides and skins project and that he may be able to provide some funding. The purpose of this was to maintain American interest in anthrax as the disease was much more prevalent in tanneries there than in woollen mills.[107] Indeed, in 1919 the British Treasury had proposed that international collaboration was necessary to solve the global problem of anthrax disinfection and without this, local disinfection stations needed to be established in Britain order to deal with imports. Yet, the idea provoked a response from the Home Office that local sanitary authorities in Britain could not possibly combat a problem which needed such an international approach. Home Office officials were committed to pursuing an international agreement on disinfection.[108] Another example of enthusiasm for international agreements was a speech at the Liverpool dinner for the International Commission on Anthrax in 1922. The Lord Mayor of Liverpool had announced that he hoped that a method for the disinfection of hides could be discovered soon, as recently another Liverpool tanner had died from the disease.[109] However, the British had been hoping for an international agreement on wool disinfection stations in locations across the world, not a corresponding research committee on anthrax in the leather trade.[110]

Delevigne commented in correspondence in 1924 that Germany and the USA were not represented on the Mixed Committee, and so he was surprised that there was an insistence that the British should have an expert even though Legge was too busy to take part.[111] As the data which was submitted to the International Anthrax Commission in 1922 revealed, anthrax in hides and skins was a much more serious problem in Germany and the USA than in Britain.[112] However, the USA did not join the ILO until 1934, even though it had sent an unofficial delegate to the Anthrax Committee in 1922. Later in the year the ILO obviously decided to take Delevigne's recommendation and appointed a representative from Germany. It was also agreed, at the behest of their governments, that Colonel Hutchinson from India and a representative from Japan could attend meetings of the Correspondence Committee.[113] Despite reluctance from the British, Butler and Thomas demanded a replacement for Legge. First Legge suggested Duckering to Albert Thomas, and second, the representative who was chosen was Edgar Collis, who has been named, along with Legge, as one of the foremost figures in transnational dangerous trades.[114] In 1915, Collis provided the Milroy Lectures for the Royal College of Physicians on the topic of silicosis.[115] Not only this, but he had also carried out research into cases of anthrax transmitted by shaving brushes.[116]

There was no lack of individuals trying to find remedies for anthrax, especially in Italy.[117] British innovators tried to offer help; the Springfield Research Laboratories in Widnes, Lancashire, for example, contacted Ludwik Rajchman, director of the Health Section of the League of Nations, asking him to look at the results of their laboratory tests regarding a 'secret discovery'.[118] Yet, by 1928 there was no foreseeable prospect of a successful way of disinfecting hides and skins without causing injury to the materials. Therefore, the correspondence committee decided to draft regulations to protect people working with hides and skins, which were finalized in 1930. In 1924, sanitary measures had already been approved in order to assist dockworkers, a group which it was resolved should be the subject of further investigation. The draft regulations which were drawn up for the hides and skins industry were very similar to the nineteenth-century rules for the wool trade. They focused on cleanliness in the workplace, personal hygiene and the provision of washing and eating areas for employees, notification of the disease and the display of information warning employees about the disease, and information about which doctor or hospital to visit for help and for the anti-anthrax serum. However, they also included mechanical means of moving the hides around the workshops where possible. The measures for protecting dockworkers included mechanization, ventilation in areas where bundles were opened, and the disinfection or destruction of wrappings. The resolutions were distributed to governments in 1931.[119]

Meanwhile, in 1924, the Trades Union Congress (TUC) protested against the dangerous goods which were unloaded in British ports, and criticized the Home Office.[120] In 1929 the work of the ILO and League of Nations anthrax committee was challenged by the TUC which discussed the topic at its General Council in Belfast, listing the risks of anthrax in the order 'hides, hair and wool'. The TUC made a resolution to ask the ILO to continue its work on anthrax, to which the ILO responded that the work had continued since 1923 and had focused on hides and skins.[121]

Although there was still no conclusion to the bacteriological experiments on hides and skins, the Mixed Committee was still hopeful about the results of one process.[122] Investigations carried on into the 1930s. After the Industrial Hygiene Committee of the ILO included the topic of anthrax and transport workers in its agenda in 1931, a questionnaire was distributed in 1932 asking about anthrax amongst workers in the transport industry, in addition to requesting information on lead poisoning in various workplaces.[123] This development revealed the continued British contribution to this discussion. Collis wrote to Carozzi to express his concern about tanneries which were 'dirty, sloppy places, which require to be kept clean and decent if civilized man is to work in them in the twentieth century'. The British government was planning welfare regulations for tanneries and Collis compared them with proposed ILO regulations. The British were also planning a code to combat possible infections from hides and skins from Africa and Asia. A first-aid code had already been established as part of the 1923 Workmen's Compensation Act and ventilation to remove dust should already have been in practice in factories. However, Britain was not progressing very far as the welfare order for individual hygiene had been suspended because of the difficulty in defining all of the different occupations in tanneries and the protective garments that each group should wear.[124] Mortimer and Melling have argued that the British had lost their enthusiasm for international collaboration by the time the ILO proposed that Special Regulations should be ratified in order to protect dockers and porters in the early 1930s and by the 1940s. Even the Liverpool disinfection station had become a 'curiosity', with its extractor fans turned off because of the noise.[125]

As a brief conclusion to these international discussions about leather, in addition to the lack of contemporary attention as a result of less union support regarding anthrax and dockworkers than for the disease among woolworkers prior to 1924, the climate of casual labour in the London docks versus skilled labour in the Bradford wool trade was crucial. This helps to explain why people working with leather in London received much less state support for such a long time. The lack of enthusiasm for the international project to find a means of disinfection provides strong evidence for continued disinterest in a bacteriological method of assisting leatherworkers. The Home Office and Treasury clearly prioritized the wool trade.

A series of inquests and publications reveals that there were even wider risks of infection with anthrax in London's workplaces. As anthrax infected so many species, work involving products derived from large animals was a risky business, for example among butchers and people working with horses.[126] As a precursor of what was to come in the twenty-first century with the anthrax postal scares, a man at the Parcels Post Office at Clerkenwell died from an infection at work in 1898. His duty had been to use untanned leather to repair the wooden packages and baskets used for the foreign service. The employee used hides which were 'simply cut and soaked in water'.[127]

Laboratory environments could be very dangerous. A laboratory attendant at University College Hospital died in 1911 even after he was given anti-anthrax serum and had an operation to remove the boil. His supervisor, Dr Francis Thiele, lecturer on bacteriology, was in charge of the University College Hospital laboratory. He had been training medical officers of health who needed to learn how to handle organisms for their new role. He believed that the only way the organisms could have escaped was if one of the students had accidentally spilt some of them on the outside of his tube and did not report that the tube needed to go straight into the steam sterilizer. The inquest recorded death by misadventure.[128] Most of these occupations were listed by Donald Hunter in his *The Diseases of Occupations*, a seminal volume on occupational health published in 1955. He also mentioned an intriguing case at London Zoo.[129] This outbreak involved humans undertaking dangerous autopsies on elephants.

Anthrax at the Zoo

In 1926, anthrax in London made the national press again with a zoonotic disease outbreak.[130] Two Indian elephants died at the Zoological Gardens in December. The first elephant to die, called Indiarani, did not receive an autopsy very quickly, so there was no trace of anthrax bacilli. Gas gangrene bacteria were found instead, possibly illustrating how the bacilli of putrefaction could overcome anthrax bacilli. The second elephant, Sundermalah, had a scab on her elbow and smears of the tissue beneath revealed enormous numbers of anthrax bacilli. The oil-cake feed was tested bacteriologically but no anthrax bacilli were found. The pig oil which had been rubbed on them was also tested and nothing was found. After Sundermalah died, all of the members of staff who had been in contact with the elephants were examined. Five men were found to have cuts, sores or abrasions. Four men contracted the disease, including one who did not have a wound. One of the other men had been highly involved at the autopsy and another helped to cart the remains away. One man only had contact with the first elephant, strongly suggesting that she had suffered from anthrax.[131]

The pathologist at the zoo, Henry Harold Scott, trained at Bart's and St Thomas' Hospitals and worked as a house physician at St Thomas' before serving

in the South African War. Subsequently he worked in the colonies as a pathologist in Jamaica, and after a stint as a pathologist for the army at Aldershot and
a fellowship in comparative pathology at the London School of Hygiene and
Tropical Medicine (LSHTM), he travelled to Hong Kong where he briefly
worked as a pathologist and bacteriologist. He returned to Britain in the early
1920s and worked for LSHTM in addition to the Zoological Society of London
and so was at the interface of human and animal health.[132] Although he was presumably distressed at the loss of two valuable animals, Scott must have relished
the opportunity to carry out research on the anthrax outbreak. While in Hong
Kong, he had carried out autopsies on 300 fatalities among Chinese labourers
and he compared his findings with cases of tuberculosis among animals at the
zoo, including birds and primates. His 270-page study was published as a report
for the Medical Research Council in 1930 but he began publishing this work in
1927, at a similar time to the anthrax outbreak.[133]

Cases of anthrax at the zoo provided another opportunity to research a bacterial disease which crossed species. Indeed, Scott devoted half of his report on
the outbreak to the elephants and half to the 'human aspect'. He suspected that
anthrax could very well appear among the people who had cared for the elephants in life and death and he took this opportunity to research how malignant
pustules first arose in humans. Scott reported that he looked out for spots, with
assistance from other members of staff at the zoo, meaning that he had studied the exact symptoms and size of the pustules as they occurred. This defined
the early signs of anthrax in humans and the lag in the appearance of a pustule
following exposure to the bacilli. He argued that the period of incubation was
six or seven days, challenging the current beliefs that it was between four hours
and three to four days. He concluded that a wound was not necessary for bacilli
to infect the skin because a hair follicle could become infected. Therefore arms
as well as hands needed to be thoroughly washed following autopsies. Finally
he proposed that the pustule began as a very small, painless, insignificant spot
rather than the 'small red nodule surrounded by reddish-blue blisters' containing
fluid, as indicated in the textbooks.[134]

The Ministry of Health was involved in examining the outbreak. They determined that all necessary steps had been taken to prevent further disease. Despite
the no fault clause in Workmen's Compensation, no discussion of compensation
for the workers involved is documented in the minutes, but the zoo's council
noted that in addition to losing two elephants, there would be considerable
expense for the treatment of the human cases and for watching contacts, as well
as for the destruction of contaminated material.[135]

Risks of anthrax in the London workplace were wide-ranging and news-
worthy, disseminating knowledge of the disease to the broad readership of the
national press. However, in 1915, a problem with shaving brushes had already

occurred which would have alerted much of the male population to the dangers of anthrax in London and across Britain, as well as the East Coast of North America and Australasia.

Anthrax in the Wider Community: Shaving Brushes and the First World War

In June 1915 a solicitor's clerk was given a shaving brush as a present. He had only used it a couple of times before he noticed a lesion on his neck. After three days he went to the West London Hospital where he died on the following day. Reginald Elworthy, the pathologist at the hospital, tested the patient's cheap shaving brush, which was made of imitation badger hair. After Elworthy discovered virulent anthrax spores, he tested more brushes bought from the same shop and found more spores. Another case appeared at the same time and was traced to a brush from a different shop. A hawker in Deptford developed a lesion within a week of purchasing a new shaving brush. His lesion developed much more slowly and after reporting to the London Hospital, he recovered after a week, four weeks after first noticing the problem. All the brushes originated from one dealer and one factory. Elworthy supported the evidence of his bacteriological examinations with animal inoculations.[136] On referral to the LGB, the topic was investigated by Dr Francis Coutts in collaboration with Edgar Collis of the Factory Department of the Home Office. The hair from which the brushes were made was determined to be horsehair, but it had not been disinfected according to Home Office regulations as the hair which the manufacturer purchased was labelled as goats' hair. The manufacturer promised to disinfect all hair in the future. In total nine people had been infected by this manufacturer's brushes. The remaining brushes had been spread far and wide, nationally and internationally. London County Council and the LGB had to trace all of the distributors and retailers in Britain, including Ireland, and the MOH for London had to inform the High Commissioners for Australia and New Zealand. Despite this, references to the manufacturer, wholesaler and individual retailers were anonymized in Coutts's report.[137] Other brushes involved in the incidents came from England, Canada, New York and Japan and again distributors and retailers had to be traced. For example, 30,000 brushes were recovered from the 43,200 brushes which had been imported from Japan. The Japanese brushes carried such a high level of infection that by 1917, the LGB had contacted the Foreign Office to ask the Japanese government to ensure that hair coming into direct contact with humans was disinfected.[138]

There were six cases in 1915, thirteen in 1916, fourteen in 1919, thirteen in 1920 and three in 1921. In total there were eighteen deaths. Heightened attention to these cases may have occurred as they struck a range of people; victims included a greaser, an electrician, a gunpowder-works labourer, a clergyman and the solicitor's clerk. The problem also directly hindered the war effort in a small way. Among the troops the problem was more concentrated than within the civilian population; between 1914 and February 1917 there were eighteen cases in England, with four definitely caused by shaving brushes. Among the British troops in France there were twenty-eight cases. There was no proof of origin but the soldiers had all been issued with shaving brushes as part of their equipment. There were a further six cases in the Navy between December 1919 and January 1921.[139]

Therefore, the contamination of shaving brushes with bacteria became a national problem, resulting from international trade. Locations in which people were infected by the shaving brushes included the south of England, namely in Torquay (brush purchased in Bath), Eastbourne and Folkestone; the Midlands, namely in Coventry and Birmingham; and the north, namely in Sheffield, Liverpool (purchased in Canada), Mossley near Manchester (purchased in Blackpool) and Halifax (purchased in London). In Scotland cases occurred in Glasgow, Leslie in Fife (purchased in Glasgow), and in Dumfries. These incidences involved brushes being sent from across the country for bacteriological testing by Arthur Eastwood, the bacteriologist to the LGB, with some tested locally, for example by Eurich in Bradford, or by the Liverpool bacteriologist and professor of pathology at the university, J. Martin Beattie.[140] Although the brush involved in the Liverpool case was bought in Canada and manufactured in New York, the incident inspired the MOH, Edward William Hope, and Beattie to investigate brushes sold in the city. Of nine different sorts of brush, four types were found to include some infected with anthrax spores. One case was confusing as there had been no initial lesion through which the bacilli could enter.[141] As we have seen, the investigations at London Zoo clarified that infection could only require a hair follicle.

The flurry of cases which occurred at this time was as a result of the war. There was a cessation of supplies of cheap shaving brushes which had formerly been imported from Germany. This resulted in a wholesaler acquiring the goat's hair and sending it to a British manufacturer. The stock of hair was actually two years old as there had been little demand for such cheap hair prior to the war. The consignments were known to be mixed with horsehair, but the wholesaler did not inform the manufacturer. Coutts was informed that prior to the war, this Chinese hair was suspected to carry anthrax and that in 1899 manufacturers had agreed with the Chief Inspector of Factories and Workshops not to buy horsehair from Russia, Siberia or China as there was no guarantee that it had been disinfected. As the manufacturer had been sent invoices clearly stating 'goat's hair', there could not be any proceedings against him for breaching the Home Office Regulations by not disinfecting the hair.[142]

The Times disseminated information about the risk. In 1916, readers were warned about a consignment of infected cheap Japanese shaving brushes that had arrived in Newcastle – small ones with black japanned wooden handles and a thin brush of white hair. Medical Officers of Health had been warned and were recovering the brushes in Newcastle and the Midlands.[143] However, Japan was not the only source, with some infected brushes arriving from New York in the same year.[144] In 1920, complex instructions were printed in *The Times* regarding how to clean a shaving brush before use: the hair of the brush must be washed in soap, warm water and washing soda, then rinsed in warm water, before immersion of the brush in a disinfecting solution of two tablespoons of formalin in half a pint of water at a temperature slightly above body heat. The person cleaning the brush needed to make sure that he did not touch the hair of the brush with his hands before this procedure had been completed.[145]

Hamer, MOH for London County Council, argued that more stringent regulations needed to be imposed on imported hides and fibres for brushes.[146] In February 1920 the Anthrax Prevention (Shaving Brushes) Order was issued to restrict the importation of Japanese shaving brushes as a result of the number of random samples which had been found to be infected. Furthermore, there was still concern about brushes which had already been imported – it was thought that a considerable number were still with retailers. Instructions were issued that brushes should be disinfected in formaldehyde before use.[147] Concerns regarding imported brushes led to discussion as to what formal regulations could be brought in to sterilize the brushes so that they were the same standard as those made in Britain.[148]

In 1921, the Ministry of Health issued a circular about the danger of shaving brushes and anthrax. The medical correspondent of *The Times* again incited fear among the papers' readers by arguing there were still a 'large number' of infected shaving brushes at large. The disinfecting remedy was published in the newspaper again and with a warning that complete disinfection was impossible as the spores could be imbedded within the brush. The article also listed, in a table, the rate of success of various remedies among 800 cases of anthrax caused by a variety of trades and agriculture, in addition to shaving brushes. Without treatment this table showed that the case mortality rate was 48.3 per cent and that even with treatment, including the excision of the pustule and/or anti-anthrax serum, the rate was between 4 and 14.4 per cent, with serum alone providing the best results.[149] However, the statistics presented are not contextualized and do not account for the length of time between infection and seeking help, or the portal of infection for these 800 cases. When Professor Vilhelm Ellerman, a senior pathologist at Bispebjerg Hospital and the University of Copenhagen, died from an anthrax-infected shaving brush in 1925, a reader wrote to the *BMJ* to argue that he advised his friends 'never to buy any shaving-brush, but to use instead a small red rubber sponge, which, with good shaving-soap, is as good as a brush, costs far less, and above all, is safe'.[150] With a total of twenty-one deaths between the first years of the First World War and the early 1920s, there was a significant response from bacteriologists, the press and from the government, including the Foreign Office as well as the Ministry of Health. The fear that any man who shaved could be struck by the disease if a shaving brush was made from imitation badger hair surely triggered a much faster and more extensive approach than the attention paid to anthrax within the leather industry.

Conclusion: Locating Anthrax in Cities, Workplaces and Homes

In comparison to Liverpool, workers and the medical profession were far more aware of anthrax in London. This was partly as a result of the medical involvement at Guy's Hospital. Crucially, as a result of the interest of a physician and the bacteriologists at Guy's, the affliction of a malignant pustule and the possibility of a subsequent systemic infection within the leather trade was completely redefined. No longer was the condition thought to arise from chemical poisoning from the substances which were used to prevent putrefaction of hides and skins. However, in contrast to Bermondsey, Bradford was a highly politicized city, where the Independent Labour Party was born and woolsorters were skilled, intelligent workers who were relatively highly paid. Although some dockers belonged to unions, the National Woolsorters' Union was much more focused on solving the problem of anthrax infections, for example insisting on specific interventions such as the temperature of the workroom and fan speeds for ventilation.[151]

The success in finding methods of disinfecting wool, in which Britain was a world leader, may be a key reason as to why historians have focused on the woollen industry rather than leather. A solution was never found which could disinfect leather without damaging it. Therefore there is no trajectory towards success. However, during the 1930s, antimicrobial sulphur drugs were developed which superseded sera in treating infected patients. Subsequently, in the 1940s the antibiotic streptomycin was found to successfully treat anthrax. By the 1950s, human vaccines were widely available which were used for the protection of workers in industry.[152] This reduced the incentive to find a means of disinfection for leather. Yet, this problem has not been forgotten within the trade. In Bangladesh an outbreak resulting from leather resulted in illness among at least 600 people in 2010.[153] Cases of the disease have also continued through agriculture, butchery and consumption of meat. For example, an epidemic of 10,000 human cases occurred in Southern Rhodesia in 1979–80. There are still between 10,000 and 100,000 human cases of anthrax annually across the globe, particularly in Chad, Ethiopia, Zambia, Zimbabwe and India.[154] Historians have also commented on present-day risks from leather. An American died from playing an imported drum as the skin was contaminated with anthrax spores.[155] In noting the risks for tanners, Carolyn Steedman has pointed out that historians are in danger. She speculates whether cases of 'brain fever' amongst scholars, which occurred nearly 200 years ago, could be as a result of anthrax spores in the leather binding of books.[156]

Chapters 1 and 3 have revealed fast engagement with the introduction of the concepts of bacteriology in the hospital and in industry. Yet, despite the immediate use of the laboratory in order to understand anthrax in London by Guy's Hospital and the LGB, there was nowhere near as much publicity about the disease in London as in Bradford. Examining the government responses to the disease until the early 1930s reveals a continued lack of interest in solving problems within the leather industry. Waldo, the coroner for Southwark, appears to have made public proclamations about the problems which dock labourers and leather workers faced, yet the doctors at Guy's do not appear to have publicly campaigned about the disease in the way that Bell did in Bradford, even though they spent a great deal of time trying to find therapies, disinfectants and antiseptics. As the problem of anthrax in leather failed to gain enthusiastic support from the government, and scientists have not succeeded in finding a way to disinfect hides without damaging them, these issues have not previously attracted much attention from historians who have been intrigued by the response to anthrax in Bradford. Therefore, the first half of this chapter is in contrast to the previous chapters in showing that bacteriology was of localized use within Guy's Hospital, but of much less use as part of workplace negotiations until the 1920s.

The latter half of the chapter reveals a much more proactive and public response to the bacterial disease. By the 1910s and early 1920s, the national press and the Ministry of Health were disseminating instructions on how to rid shaving brushes of bacilli, assuming a public understanding of the bacterial causality of anthrax. The following chapter continues the cautious tone of the present one in addressing the early reception of bacteriology in the 1880s in the wider community. The chapter weighs up whether bacteriology influenced public responses to disease or whether communities were more informed by everyday hygiene, sanitation and epidemiology.

5 CHARITY, COMPENSATION AND CARRIERS: THE DEVELOPMENT OF BLAME AND RESPONSIBILITY IN RESPONSE TO EPIDEMICS

> But disease germs are not our worst enemies. A new danger has arisen: an insidi-
> ous enemy who works without ceasing, whose names are fear, rumour, false report,
> folly, ignorance, cowardice, and others equally unpleasant. This enemy has seemingly
> seized upon the minds of many of our once friendly and kindhearted neighbours in
> outlying parishes, so far destroying their good nature and commonsense that they are
> now actively engaged in boycotting Malton and its too-hard-hit tradespeople! Our
> neighbours are kicking us when we are down.[1]
>
> E. Tracey Archer, *Malton Messenger*, 19 November 1932

Tracey Archer's concerns are indicative of the discussion which follows; although
bacteria were generally acknowledged in popular responses, reactions to epidemics were largely informed by concepts which had a much longer history than
bacteriology, such as fear of and flight from afflicted towns. Epidemics of disease
generate a wealth of sources for historians to explore but the outbreak of typhoid
in Malton, Yorkshire, in 1932 resulted in more personal accounts from individuals than most: several boxes of correspondence and press cuttings gathered by
the Typhoid Relief Fund, revealing the suffering, concepts of disease and beliefs
in responsibility for ill health among this local community.

The historiography of epidemics of diseases such as cholera, typhoid and
typhus has examined public reactions to disease: flight, panic, sanitary policies,
cessation of public gatherings, religious ceremonies, charity, riots and violence,
and suspicions of people in authority and of foreigners.[2] This chapter explores
most of these themes, especially inquiring whether they illustrate knowledge
of the new science of bacteriology and how this insight influenced reactions to
epidemics. A major thrust of the chapter is to question whether the significant
development of blame and responsibility for typhoid epidemics between 1882
and 1936 stems from confirmation of how disease was transmitted by bacteria
and by human carriers of the disease. Linking with the following chapter which
focuses upon criticism of the local medical profession and of local government
in Croydon in 1937–8, the following case studies explore how victims of disease

and their families became increasingly litigious, mirroring the reactions in the workplace which have been illustrated in Chapter 3.

The focal points of this chapter are an epidemic of typhoid in Bangor, North Wales in 1882 (690 cases, including 89 deaths from a population of about 9,000), and in Malton, North Yorkshire, in 1932 (249 cases, including 23 deaths from a population of 4,500, as well as 21 more cases and 1 death outside the town).[3] However, in order to connect these epidemics together across a gap of fifty years, a variety of typhoid epidemics and one diphtheria epidemic are briefly explored, in particular in Marlborough, Wiltshire, in 1890; in Maidstone, Kent, in 1897; in Eccles near Manchester in 1910; and moving forward to Chapter 6, in Bournemouth, Dorset, in 1936.

The decision to investigate the epidemic in Bangor is comparable to the choice of anthrax in Bradford in Chapter 3, with the typhoid bacillus only just having been isolated. Diphtheria at Marlborough College is explored because an invaluable collection of letters from parents has survived, not only for the outbreak in 1890 but for a previous outbreak in the 1850s, allowing a comparison of how ideas changed over time in the same location and with the same means of communication. Typhoid and diphtheria have both been classified as 'house diseases', with typhoid thought to be spread by faeces in the water or air and diphtheria thought to be particularly spread by 'foul air or sewer gas'.[4] The epidemic in Malton also generated a wealth of primary sources, the town facing considerable financial consequences for trade, with the rural town's commerce already in a vulnerable position owing to the Great Depression. The chapter, therefore, focuses on the same time period as Chapters 3 and 4, before pushing the wider narrative forwards into the final substantive chapter which explores the extraordinary response to a typhoid epidemic in Croydon in the late 1930s. To provide background for the study of these epidemics, it is necessary to explain the general symptoms of typhoid sufferers and to explore the ways in which the disease has captured the attention of historians.

Victims of typhoid fever suffer from diarrhoea, abdominal pain, a temperature, headache, cough, exhaustion and sometimes red patches on the abdomen. The disease can lead to pneumonia, and when fatal, is preceded by gastrointestinal haemorrhage and coma. The typhoid bacillus was discovered in 1880–1, and isolated in 1884. In 1896, the Widal test was developed to test for antibodies in the blood. A vaccine was produced in 1898, and in 1906 the concept of the healthy carrier was proven.[5] The mortality rate was 10 to 20 per cent of sufferers before the advent of antibiotics, to which typhoid was sensitive, in 1948. The disease can last for weeks. A total of 3 per cent of people who transmit the disease show no symptoms and become carriers. Methods of transmission include faecal-oral, for example via faeces on unwashed hands, contaminated water supplies and flies.[6] When bacteriology influenced public knowledge of the means of transmission of typhoid, it led to feelings of disgust, as victims knew they had consumed a little excrement from another person.[7] According to William Budd, writing in 1873, typhoid fever took at least 15,000 lives in the United Kingdom each year, with nine or ten times as many cases of the disease.[8] The disease killed more people than cholera in the nineteenth century.[9] There is a general consensus that typhoid is an excellent indicator of the extent of the success of public health policy.[10] Previous studies of typhoid outbreaks have focused on local and international politics, public health issues such as purity of water or food, and the topic of carriers inspired by the famous case of 'Typhoid Mary' Mallon.[11]

Several themes will run through the three detailed case studies within this chapter, and permeate the briefer accounts of epidemics. First, reactions which preceded the bacteriological era will be compared and contrasted; these include religious beliefs, reactions to miasma and sewer gas, sanitarian beliefs regarding water supply, and reactions to public gatherings and flight from epidemics. Second, the influence of bacteriological ideas will be identified. Were communities eager for the results of bacteriological tests and how did they think they would help to stall an epidemic? How did the concept of carriers affect communities' responses to epidemic disease? To test these transitions, a major theme of this chapter and the next is the increasing trend towards attributing blame and responsibility for bacterial disease.

Bangor, 1882

The epidemic of typhoid fever in Bangor allows a comparison between the reception of bacteriological knowledge in the workplace, studied in Chapters 3 and 4, and the experience within a local community in the early 1880s. Did the Bangor epidemic result in as quick a dissemination of bacteriological knowledge as a much more limited outbreak of bacterial disease within West Yorkshire workplaces?

In 1882, the *Carnarvon and Denbigh Herald* published a story of a doctor from Dublin who had visited Bangor a couple of months before the epidemic, asking whether illness was present. On hearing it was not, he exclaimed, 'you will soon have, for I observe typhoid poison in the air'.[12] During the epidemic, Dr Rees, the MOH, proposed acidulated solutions which could tackle typhoid germs in both sewage and sewer gas.[13] Ten years later, a doctor called Emyr Price wrote a thesis on the epidemic, and still he speculated that some of the cases were caused by sewer gas.[14] Although pathogenic bacilli had been discovered by this time, bacteriology was not yet a fully established discipline. Therefore, the response of these medical practitioners supports the pervasive arguments made by Michael Worboys and Nancy Tomes that the advent of bacteriology did not result in a sea change in theories of disease.[15] Rees and Price were probably 'contingent-contagionists', combining the idea of infection through poisonous air

which was capable of contaminating the water supply, with individually transmitted contagion. This trend was common in the 1870s and 1880s and even if
doctors believed in one mode or the other, they generally advocated protection
from both sources when communicating with the public.[16]

The concept of sewer gas was a development of the idea that bad drains and
decaying matter would emit bad odours which, along with general cleanliness,
were theories associated with the sanitarian movement which was at its height
in the 1840s.[17] Although sewerage systems had been essential in improving environments, from the 1870s, the gases they were thought to produce within the
home resulted in pipes, sinks and toilets becoming a source of fear. In order to
combat the problem cellars needed to be dry, indoor water pipes securely fitted
and all drains needed traps to stop sewer gas flowing back into the house. A long
pipe would direct the gas up the side of the house and over the roof.[18] 'Sewer
gas' was a topic which still warranted much attention in the early bacteriological
age, even in a hospital setting. Indeed, in 1883 in the Addenbrooke's Hospital's
weekly minutes the following was recorded:

> In the central block the soil pipe from the Matron's closet was found leaky. The soil
> pipe from the closets adjoining the nurses cubicles permitted a free discharge into
> closet + chamber and thence to staircase and corridor of odorous test smoke intro-
> duced at the bottom of the pipe. This was found to be due to the drying up of the
> water in the trap to the closet pan from disease. The bell trap drain in the beer cellar
> which joins this system is liable to dry up also from disuse and permits the entrance
> of sewer gas within the building.[19]

References to sewer gas continue in the Addenbrooke's Hospital records until
at least 1887.[20]

In addition to drawing on beliefs from the 1870s and before, the medical
profession considered new theories during the 1882 epidemic. Although bacteriology was of no use in providing any proof of the origin of the epidemic, as
would perhaps be expected so early in its history, Rees announced at a meeting of
the local board that he was optimistic about its future use as the typhoid bacillus
had only just been discovered.[21] Indeed, it was reported in the *Carnarvon and
Denbigh Herald* that Rees had previously said that the typhoid bacillus had not
yet been identified by researchers, but wished to 'modify' this statement as he had
found reference to the discovery in a medical journal. He read out portions of
the article to the city board in Bangor. However, Rees stated that typhoid bacilli
would not be found in Bangor water, even though he thought that they were
the cause: 'it was such an infinitesimal germ got into a large body of water. They
might put a jug 50 times under the water and get nothing, and perhaps get it the
51st time'.[22] The local press agreed with the city's MP, William Rathbone, that
'notwithstanding all the recent discoveries of science much more has yet to be

explained with reference to the origin of typhoid fever and its long continuance in a locality where it has once manifested its unwelcome presence'.[23] The discovery of the typhoid bacillus did not generate the same interest as the discovery of the anthrax bacillus, perhaps because anthrax was the first bacteria to be established as the specific cause of disease, and it was clear that this was the case due to its size and life cycle. Eberth's and Klebs's isolation of the typhoid bacillus in 1881 probably did not make a significant public splash as Klein had already incorrectly claimed to have discovered the microorganism in 1874.

Despite the brand-new knowledge about the bacterial origins of typhoid fever, members of the city administration chose to ignore the MOH and most of the doctors, as they believed they knew the cause of the epidemic.[24] The MOH and the medical profession were at loggerheads with the city's board, resulting in politely abusive letters in the local press and the *Lancet* regarding who was right over the course of action regarding the epidemic.[25] An article in the local newspaper showed concern at this lack of regard for expertise, stating

> It is but seldom that the reports of a district medical officer of health are scanned with a very critical eye by those before whom they are formally laid. As a rule there are but few persons bold enough to challenge the official statements made by a medical man, and the conclusions at which he has arrived.[26]

Criticism of the doctors' position on the aetiology of the disease continued when Florence Nightingale was contacted for advice by Rathbone. Rathbone and Nightingale had collaborated to plan workhouse and district nursing in Liverpool in the 1860s and so it was natural that Rathbone would call upon the woman who had been so influential in improving hygiene within armies and hospitals. Nightingale ignored the ideas of the 'Medicos', siding with the administration, and focused on the sanitarian approach that engineers were necessary in order to solve problems with pipes and drains.[27] Having studied sanitation in India in great detail, though never visiting the country, Nightingale doubted that typhoid fever was even spread in water, as not a single case in India had been traced to 'enteric fever poison' in water and there were no sewers there to create sewer gas.[28] Nightingale has been commonly portrayed as rejecting the bacterial aetiology of diseases, as this was too random a mode of infection for her, and contradicted her beliefs in individual morality in preventing disease, and that filth or stagnant air could be controlled.[29] As with many members of the health-care professions, Nightingale's ideas were gradually evolving: in the 1860s she and her physician, John Sutherland, had written about microscopically examining sediment from water. So her comment on typhoid not being spread by water might be empirically based and only in reference to that disease. She was gradually convinced about bacteriology as, by 1891, she advocated that slides illustrating germs should be shown in village demonstrations in India.[30] In any case, while the MOH in Bangor wanted to find the cause of the bacilli in the water supply, the sanitarian approach to fixing pipes and drains may have assisted with that goal, although it was not useful in proving the source of the epidemic.

In order to support those suffering from the disease, a charitable fund was established, with subscriptions recorded regularly in the local press.[31] The Bishop of Bangor made a large donation, and also offered his land for hospital tents, allaying panic in other parts of town where hospitals had been set up.[32] Ladies volunteered their labour.[33]

Along with this charitable response, from a lay perspective responses to public gatherings and flight from the city are the most notable reactions to the epidemic. Studies of cholera have shown that in Russia in 1830, people did not attend a large annual fair when an epidemic was looming, whereas after the introduction of bacteriology in late nineteenth-century Naples, people were not afraid to congregate, especially for church ceremonies and fairs.[34] By studying school logbooks which include comments on local events within the town, a similar response can be found in Bangor in the same period. Glanogwen National Girls School closed from 21 August to 25 August 1882 because of the epidemic and also due to the Church eisteddfod, a traditional Welsh festival of singing and poetry with an element of competition. Holidays were also given at various other schools during the epidemic for different large fairs and an eisteddfod at the end of October.[35] Church festivals and services continued to attract large crowds. The local newspaper and school logbooks record annual harvest thanksgiving services taking place in all the chapels in Bangor during October, even though in the same newspaper article it was reported that fresh cases of typhoid were still appearing.[36]

In complete contrast to these public gatherings, other people responded by fleeing the epidemic. In Bangor, many of the schools shut over the summer for a longer time than the normal vacation period. One of the head teachers thought the low attendance at the schools was as a result of panic rather than illness.[37] Another school investigated the low attendance of forty on 14 August and discovered that 'several children were attacked, but many more were kept from School through fear'. On 25 September, when the school reopened again, the low attendance of 70 out of 135 was explained: 'Many boys still sick with Fever, or recovering from it. Several boys with their parents have left the city ... others have been sent away until fever abates.'[38] The public response was therefore divided between some people who carried on as usual and others who sought to escape. Public, administrative and medical responses in Bangor were as expected at a time of flux with regard to the new evidence that typhoid was spread by bacteria.

Marlborough, 1890

Letters regarding diphtheria which were written to Marlborough College in 1858 and 1890 reveal a stark contrast between faith in religion and the temptation to withdraw boys from school and discussion of drains and water analysis. In 1858, some parents expressed complete faith in the school, or in God. One wrote,

> I regret that Diphtheria has occurred, and with fatal result in one instance: but I have no apprehension about my son, beyond which I ought to feel wherever he may be placed – I am perfectly satisfied that all due precaution + all necessary care will be taken – and I desire to leave my child in the Lord's hands.[39]

Another wrote,

> Not being at home there was some delay in my receiving yr circular with respect to the Diphtheria. I feel full confidence that every precaution, + every possible means will be used, trust that thro' God's blessing they will be effectual.

However, it is worth noting that the main occupation of the children's fathers was the priesthood. Yet, even though the school was still mainly for the children of the clergy, there was no mention of religion in letters about diphtheria written in 1890, with more critical questions being asked. The lack of discussion about religion perhaps shows the secularization of beliefs about disease even within a professionally religious community, in line with studies which show a decline in religion in Britain from 1880, following a fervent religious milieu in Britain in the earlier Victorian age.[40] Charles Rosenberg has demonstrated the same transition through cholera epidemics in New York in the nineteenth century. In the 1830s and 1840s, people believed the pious would be safe, and gathered together for prayer days. However, the belief in God's influence on earthly events was already declining, and by the 1860s, ideas of avoidance of sin had given way to a concentration on hygiene and disinfection in a more secularized society.[41] Therefore, the changes in religious beliefs and the secularization of society which happened prior to the bacteriological age may have already changed attitudes about disease prevention and responsibility for disease.

The comparison of the two diphtheria outbreaks at Marlborough College shows a heightened withdrawal of boys from the school in the 1890s. After one pupil died in 1858, a mother wrote the next day to say that she was going to send for her children, as they were her eldest boys and, 'situated' as she was, she could not do 'anything, which humanly speaking, might endanger the life of either of them'.[42] However, she wrote again four days later, and was grateful she had not pulled the boys out of school, so presumably either a separate letter or a circular which was issued by the school stating the other cases were only 'sore throats' managed to calm her down.[43] Another parent wrote that a note had been received from their son stating that one of his companions had died. This boy apparently told his parents, in his mother's or father's words, that 'it was reported all boys were to be sent home as the Doctor thought it better to do so as many Boys were in Sick house suffering from same as the one died of'. This child's parents were just anxious for more information.[44] Fears in Marlborough did not reach anywhere near the proportions they did at Uppingham School in the 1870s, where a typhoid epidemic resulted in the whole school being evacuated from Rutland to Wales.[45]

By the time of the outbreak in 1890, fears seem to have increased at Marlborough. There were only seven cases of diphtheria in the outbreak, but it was reported in the *BMJ* due to the level of panic.[46] From a total of 600 students, 100 boys were withdrawn from the school.[47] As in the Bangor schools, it was difficult to persuade some of these children to return. The Master wrote to the parents on 28 October,

> I have received several inquiries from parents who wish to know when their sons should return; and, for the sake of all concerned, it is plainly desirable that absence should not be unduly prolonged. Members of the College are therefore expected to return not later than Saturday, Nov. 1st.[48]

However, it is revealed in a letter of 6 November that not all the boys had returned despite no fresh cases occurring since 17 October.[49]

A couple of parents questioned investigations into the diphtheria outbreak, though without any sense of accusation. One asked about the water analysis, suggesting all parents would want to know the results of the test, and another inquired as to whether the cause had been found.[50] The analysis indicated could, however, have been chemical rather than bacteriological. Another parent suggested a possible course of action involving classifying the students, for example by their school year, to discover in which areas of the school they had been, as drains had been the cause at Haileybury School.[51] Haileybury School in Hertfordshire had suffered two deaths from diphtheria in 1888.[52] The discussion about drains reveals that this parent probably thought that 'sewer gas' was the cause of diphtheria. Although this demonstrates continuity, the most remarkable change between the 1850s and 1890s was the decline in religious ideas of disease and the increased number of absences from the school during the later epidemic. This strongly suggests that scientific knowledge of disease stimulated fear. In her study of typhoid at Cornell University in 1903, Heather Prescott deduces that increased knowledge of disease transmission brought more fear, so perhaps knowledge of bacteria encouraged people to fly from Marlborough.[53] However, although parents were frightened and wanted to know more about what was happening at Marlborough College in 1890 regarding the diphtheria outbreak, there was no sense of blaming the school in their letters.

Maidstone, 1897

The Maidstone typhoid epidemic of September 1897 resulted in over 1,300 cases of the disease.[54] Nearly 200 cases were diagnosed within the first eight days. Immediate advice on preventing illness and references to bacteria were provided in the first announcements about the disease; the first notice of the epidemic in the press advised people to boil their water and milk, claimed that the water company were doing everything they could and that samples had been sent to London for bacteriological analysis.[55] Early in the epidemic, the *Kent Messenger* told its readers that there was no danger in entering the houses of those who were unwell, or entering the town if 'ordinary sanitary precautions' were observed and that water and milk needed to be boiled for ten minutes before consumption.[56] Several weeks later, fifteen doctors signed a letter to the *South Eastern Gazette* telling the public that it was safe to shop in Maidstone.[57] Despite this, St Paul's Schools were closed for the hopping holidays on 20 August 1897 and did not reopen until 3 January 1898, with the reopening of the infant school postponed indefinitely.[58]

Although the press dispersed information quickly, 'A Sufferer' wrote to the *Kent Messenger* to complain that the handbills of advice from the Sanitary Commission had been printed on 15 September, but had not been delivered until six days later. In addition to this complaint, he or she argued that the water rates were expensive and as such the public had the 'right to the real facts'; the Water Company had 'done well for themselves' while those suffering had 'anxiety' in addition to 'long bills' for medical expenses.[59]

The emphasis on bacteriology continued in the press, with the assumption that the people would understand the importance of laboratory tests. As soon as the epidemic was publicized by the *Kent Messenger*, mention was made of the necessity for bacteriological tests, but that chemical tests should also be sought as the results would be faster.[60] A letter to the editor of the *Kent Messenger* simply asked how often the bacteriological tests were carried out.[61] The *South Eastern Gazette* had already argued that the 'pack of parsimonious noodles', otherwise known as the municipal representatives, had decided to cut the council's expenditure and reduce tests of the water supply from weekly to once a quarter. Therefore, before the September epidemic, the water had not been tested since 25 June.[62] Yet, Alderman Spencer argued that the water tests became monthly rather than weekly.[63] A letter to the editor of the *South Eastern Gazette* argued that regular analyses would give a 'sense of security never warranted' and that filtration was necessary. Indeed, historian Christopher Hamlin notes that there were worries in the late nineteenth century that 'false security' could come from bacteriological testing.[64] Filtration had previously been rejected by ratepayers as it would have cost £195,000. Spencer complained that one of the problems was that the water supply was run privately, not by the town council.[65] However, this distance between the management of the water supply and the town council meant that the mayor and councillors could meet with the water company to complain on behalf of the citizens of Maidstone.[66]

Finally, in January, the results of bacteriological tests, 'so jealously guarded' since September, were 'at last public property'. The *South Eastern Gazette* had obtained a copy, to 'give to our readers all the principal facts', though the paper apologized that they were 'lengthy and technical' before proceeding to print a long list of when and where samples were taken. The paper explained that the test results were taken on 19 September but that because there was an average of fourteen days' incubation period, and patients took around one or two days to seek help, the typhoid bacillus had not been discovered. However, animal excreta had been discovered in the Tutsham-in-field spring and there was some chemical evidence of pollution.[67] There were rumours that the epidemic had been brought to Maidstone by transient hoppers from London, which were substantiated by the MOH. These hoppers had suffered from diarrhoea, a fact which was known because a local shopkeeper told the deputy MOH that they had frequently asked her for diarrhoea mixture.[68]

Although the immediate discussion of bacteriology distinguishes this epidemic from Bangor and Marlborough, there was a letter to an editor about sewer gas, highlighting continuity in concepts of typhoid transmission. The correspondent was concerned about the ventilation gratings opening from the main sewers onto the streets. He did not believe the assurance that they were 'perfectly healthy'. He argued that the gentlemen who were elected to represent the town's affairs should not be 'carried away by the volubility of every theorist that tries to force his ideas in the face of practicability and common sense, only to realise, when too late, the extent of their error'.[69] The same person wrote again a week later to complain of this 'great danger'. He appears to have been angry that the council dismissed his comments by stating that they had poured carbolic acid through the grates: 'Let anyone pour half a gallon of acid down one of these stench centres and 30 minutes later try to trace it'. He argued that the ventilators needed to be disinfected by 'suspended wire baskets containing charcoal and deodoriser'.[70] At Holy Trinity Church, the Reverend T. W. Mylne gave a sermon saying that prayers would 'not prevent sewer gas from entering your house if you take no precautions to keep it out'; prayers would not help if the reservoir did not contain clean water. Yet, he asked his congregation to pray for the sufferers and for advice.[71] As a result of the LGB investigation, there was a dispute over the chemical and bacteriological evidence and as to whether typhoid could be transmitted by vapours.[72] On reflection, a month later the *Kent Messenger* argued that the dispute between scientists had led to questions as to whether they could be trusted by the public.[73] Therefore, debates about noxious air versus waterborne bacteria continued among the medical profession in the very late nineteenth century.

During the epidemic, charitable relief prevailed for those who suffered, rather than retribution. The Urban District Council Special Committee, which was appointed to oversee the typhoid outbreak, requested donations of old linen, blankets, flannels, towels and sheets, in addition to accommodation for the extra nurses who had been brought into the town.[74] Nearly £28,000 was raised, mostly by the *Kent Messenger*. On 2 January, there was a service of thanksgiving with the offertory donated to the benevolent fund which had been organized by the local friendly societies.[75]

In spite of this charitable spirit, isolated complaints began early in the epidemic but the local council made a mockery of a particular person for complaining:

> Alderman Spencer regretted that certain people were going about the town spreading exaggerated statements as to the epidemic. It had come to his knowledge that there was one individual especially who had been running about and greatly exciting the public mind. This person had been making the most extraordinary proposals – for instance, that a public meeting should be called to consider the matter (laughter). But what could a public meeting do?[76]

As will be shown in Chapter 6, the contrast between this response and what occurred during the typhoid epidemic in Croydon in 1937 is extraordinary. Another person wrote in early October 1897 to ask who would be paying the medical and Urban Council bills and urged for these to be passed to the water company.[77] However, another reader argued on the same day that it was 'premature' to discuss the faults at stake until health had been restored and that the authorities needed to be helped in their 'arduous duties'.[78] J. Vincent Bell of Rochester argued in a letter to the editor that the medical officer and sanitary authority needed to be investigated by the LGB if they had not fulfilled their responsibilities, but reasoned that this was in order to prevent a future epidemic rather than to interpret the causes of the present one.[79] The LGB dismissed the idea of a public inquiry, concluding that an investigation was adequate. This decision, announced at the Urban District Council (UDC) meeting, was apparently a disappointment to many people.[80] An angry letter about the 'most disgraceful period of Maidstone's existence as a Municipal Borough' which was published on 9 October, argued that the people had paid for '*poison*' and that they had a '*right to demand*, a thoroughly impartial inquiry and rigorous compensation (*not* compensation derived from the imposition of an additional rate) for pecuniary loss entailed'.[81] At another UDC meeting, Alderman Dann joined in these complaints, saying that when 'scientific men' asked the water company to turn off the supply, the company continued to supply water without disinfecting the mains immediately: 'The people outside will not be satisfied until each and all of these Directors are arraigned before a bar of justice to answer the charge of manslaughter. It is wholesale manslaughter, nothing else (applause from the public).'[82]

One reader of the *Kent Messenger* wrote in January 1898 that the water company should be paying doctor's bills rather than the charity relief fund, and wanted people to unite to seek compensation. He said that the labouring classes were being kept from discussions regarding the epidemic.[83] By February 1898, the Maidstone Friendly Society was reported to be likely to push for a test case to meet their costs for assisting the sick and for providing funeral funds. It was noted that a similar claim against a water company had been successful in Wisconsin. However, the inquiry did not resolve the issue of blame.[84] In March people met privately to decide to bring a combined action at the High Court against the water company. Apparently the company had offered £3,000 if they did not take the case to court and this was refused by the sufferers.[85] By April the *Messenger* reported that a grant of £3,000 had been given to campaigning sufferers by the water company through generosity, not through acceptance of responsibility. This is half the money the people wanted. The newspaper considered this a good result.[86] Frank Trentmann and Vanessa Taylor have argued that consumer rights in general began with complaints about water, with their study on Victorian London focusing on the value, supply and increasing need for water for personal hygiene. However, they have not noted complaints about water carrying disease.[87] The out-of-court settlement in Maidstone is significant in the history of compensation and disease and a similar attempt to claim does not appear to have occurred again until 1938, as will be discussed in Chapter 6.

The Maidstone epidemic occurred at the same time as Almroth Wright was developing the vaccine for typhoid. An article was published in the *South Eastern Gazette* advertising that the Army Medical School at Netley would vaccinate the whole population of Maidstone but that the demand must come from the public.[88] Wright tested the vaccine on the nurses and attendants at Barming Heath Asylum in Maidstone.[89] Although there was considerable discussion about bacteriology and also mention of the brand-new innovation of typhoid inoculation, the discussion of sewer gas illustrates significant continuity with older ideas. A few years later, this topic was still discussed within the LGB report of the typhoid epidemic in Lincoln in 1905. The report argued that the main causes of the disease were always the water supply, milk or 'sewerage and drainage'. 'Sewerage and drainage' were a problem because of 'sewer-air', but in this epidemic this reason had been disregarded as the first twenty-five houses in which cases had arisen were checked and there were no defective sanitary fittings.[90]

One of the primary means of advice for avoiding sickness in Maidstone was to boil water and milk. Yet, these were not brand-new bacteriological ideas. The boiling of milk to ensure purification was advised as early as the eighteenth century for infant feeding. More recently, milk had been associated with an outbreak of typhoid in Penrith in 1857.[91] It was in the 1880s that milk was redefined as a seriously 'dangerous carrier of disease', but preservatives were generally used rather than pasteurization. It was only by the 1920s that pasteurization became common practice, one reason being that pasteurization could be used to 'forgive dirt and common slovenliness'; another was that it changed the taste.[92] The boiling of water had been advocated by the sanitarian movement so was also not a new bacteriological concept.[93] However, by the early twentieth century, a new development in bacteriology, the theory of the healthy carrier, led to a radical change in the emphasis on finding the source of an outbreak, redefining responsibility.

Typhoid Carriers

The concept of healthy carriers has been described as one of the most 'publicised and transformative discoveries of bacteriology'.[94] The theory of a healthy carrier had been suggested by Loeffler in 1884 when he cultivated diphtheria bacilli from a healthy patient, and by Koch who discovered a person who carried cholera in 1893.[95] The healthy carrier concept and campaign was formalized in Germany in 1904, stemming from ideas of ambulatory cases, sometimes known as 'Parasitenträger' by Koch and others from the beginning of the century. Typhoid carriers were identified in Germany from 1906.[96] Andrew Mendelsohn has described the search for typhoid carriers in south-west Germany at this time as the 'world's most closely watched disease', with bacteriologists given permission to access a variety of records and people in order to track down the cause of epidemic disease.[97]

The first carrier to be identified in Britain was a cook working at the Brentry Home for Inebriates near Bristol, which led to a steady trickle of twenty-eight cases and two deaths between September 1906 and November 1907. She was also retrospectively linked to cases at a girls' home near Bristol, where she had previously worked. Richard McKay argues that the tone with which the MOH for Bristol and his colleague wrote about the carrier invoked blame for the disease. For example, they referred to chronic carriers as a 'dangerous class' who contaminated food with their hands after 'defecation or micturition'. The doctors presumed that the carrier had spread the disease because she had not washed her hands between using the toilet and cooking for the children.[98]

The concept of a typhoid carrier became infamous through the case of Mary Mallon in New York in the early twentieth century. As a result of her reluctance to stop her work as a cook and to accept the results of bacteriological tests, the medical profession, the city authorities and the press held her responsible for spreading disease and she was detained twice.[99] Mallon's case was not normal New York policy, but an individual case, and other carriers were retrained in different occupations and educated so as not to transmit disease.[100] Even a bakery and restaurant owner who continued to handle food although he had been told he was a carrier received only a suspended sentence in New York in 1924 rather than being detained. Mendelsohn and J. W. Leavitt have argued that his status protected him in comparison with Mallon, an Irish immigrant cook.[101] So, it seems that apart from Mallon's famous case, carriers were never really held to blame for disease. In the epidemics examined within this chapter, local authorities or suppliers were generally held responsible for the outbreaks as they were supposed to regulate the transmission of the disease through infected water, milk or food. The discovery of carriers clearly placed an onus on these people to be meticulously hygienic. Priscilla Wald has argued that with bacteriology came manuals which advocated individual responsibility for the control of disease within wider society; whereas sanitarians believed that filth and the resulting miasma spread disease, writers informed by bacteriology argued that personal hygiene was more important for many diseases, such as keeping fingers out of mouths.[102] Yet, Tomes has urged historians to recognize that personal hygiene was already important for sanitarians.[103]

The Eccles epidemic of 1910 particularly targeted schoolchildren, as it was spread by ice cream contaminated by typhoid carriers. Over the period of half a week in November, the 'familiar removal cab of the Salford Sanatorium Committee' trawled the streets removing a 'constant stream' of sufferers, including eighty on one day. Nearly a hundred children were absent from the Eccles Parish Church schools, either through illness or 'apprehensions of parents, who [did] not all realise the distinction between infectious and contagious disease', so the schools closed for a fortnight. To allay fears, all elementary school pupils who had been in contact with a person suffering from typhoid were asked to abstain from attending school. Towards the end of a newspaper article, the emphasis was placed on the means of transmission, after announcing that the Sunday school would also be closed. The reason for the closure was 'to avoid any chance of contagion which is possible only from the common use of sanitary conveniences. It is because of this and not because any suspicion attached to the drainage or sanitary arrangements of the buildings'.[104] Following recommendations on regulation and inspection of ice-cream carts, an appeal in the local press was for charity, not for compensation, requesting the 'material help of the community' where breadwinners had died.[105] The only mention of monetary compensation was for a milkman who had been falsely suspected of causing the epidemic and been asked to stop working. The man had already sold his business before negative bacteriological tests were returned and the milkman was told he could return to work.[106]

The calls for donations triggered Will Hughes to write to the local newspaper to complain that old bed linen was unhygienic, and was therefore the 'worst kind of busybody charity-mongering'. He also cited the 1875 Public Health Act which stated that the local authority may be liable to compensate for the bedding. He claimed that the MOH should 'resent' the idea as personal clothing was a 'very, well-known method for the transference of vermin and disease'. He encouraged donations of good quality new bedding that would improve the lives of the poor, but he also used the epidemic to make a broader socio-political argument that the poor needed to be given the chance to go to work so that they could provide for themselves.[107] Indeed, this epidemic occurred within months of significant political change, the People's Budget having been accepted in April 1910; the majority of letters to the local newspaper at the time of the epidemic discussed the Chancellor of the Exchequer, David Lloyd George, the Irish question, tariffs and free trade.[108]

In contrast to the later epidemics which are to be discussed, the suspected carrier, Antonio, was publicly named, perhaps because the local population had worked out for themselves how the epidemic had been caused because they had been interviewed by the MOH.[109] He may also have been named as, like Mallon, Antonio and his colleagues were outsiders, not only being Italian but also visiting Eccles from the Ancoats area of Manchester. They were asked not to return to Eccles.[110] The epidemic was first noted in the press on 25 November, and by 9 December the carrier had been named at the local town council meeting. Of the 137 cases, 115 had 'definitely' been traced to Antonio (although the LGB report listed that there were 142 cases and 108 had consumed ice cream).[111] Antonio was named in Eccles because he sold the ice cream, but the LGB report revealed that there were fourteen people involved in the ice-cream business and that six of them had positive Widal's reactions for typhoid while paratyphoid bacilli were found in the urine or faeces of five of these people.[112] Although the press attributed little blame to Antonio, his business would have been doomed. They were asked to stop making ice cream in Manchester, and having contravened this, the new ice cream was destroyed. The house in Manchester had to be watched by the local police.[113] The *Eccles and Patricroft Journal* proclaimed there would be little demand for ice cream in Eccles in the near future.[114]

Yorkshire and London in the 1920s

A significant proportion of this book has explored the regions of Yorkshire, and London and the surrounding counties. How did reactions to typhoid in these areas develop during the 1920s? In 1921, two epidemics occurred in Bolton-upon-Dearne near Rotherham in Yorkshire, revealing people's preference of taste over the cleanliness of water. When a brief typhoid outbreak occurred, the water supply was chlorinated to prevent future waterborne epidemics. Yet, five months later, in November 1921, typhoid struck again. Many of the local population did not like the taste of the chlorinated water, and they therefore used 'irregular sources', which led to typhoid in epidemic proportions.[115] For a small place this was a huge epidemic, with 397 cases and 41 deaths, including 113 cases among children under the age of ten.[116]

In Gravesend, Kent, in 1927, a small outbreak of typhoid was determined to have been caused by a well, polluted with sewage, from which water was used to wash down a dairy and the associated utensils. All of the thirty-nine sufferers had consumed milk from one supply.[117] The MOH advised all residents to boil water and milk even after the epidemic was 'in check'. He had explicitly stated at a town council meeting on 13 May that typhoid could not be contracted through bad smells. Yet at a town council meeting later in the month, Councillor McKenzie expressed a strong opinion that this could be the case, quoting 'an authority': a sanitary engineer who had listed 'drain gas, contaminated water or milk' as the means of transmission. The sanitary engineer had apparently argued that if polluted water got underneath houses, gas could be formed causing typhoid. Indeed, McKenzie had seen cesspools overflowing in Augustine Road and argued that when they were drained and lime or chalk was put in them, this resulted in 'drain gas'. The deputy mayor lamented that the MOH was not present 'to put a totally different construction on the matter', but McKenzie used this comment to criticize the doctor for not attending the meeting. Moving on, the chairman of the Health Committee thanked the medical officer and sanitary inspector for their hard work and announced that a sample of well water had been sent to Canterbury for analysis.[118] An article in the local newspaper argued that a 'main drainage system' was needed, but the people were apathetic and were reluctant to spend money.[119]

A 'somewhat extensive' outbreak of paratyphoid fever struck London, Surrey, Middlesex, Essex, Hertfordshire and Kent in late July and August 1928.[120] However, it was discovered that cases had occurred since 8 July.[121] The *Evening Standard* sensationally referred to the hunt for the possible carrier as the 'Scarlet Pimpernel Germ Drama' and '[a] relentless germ-hunt – one of the most important investigations in the recent history of science'.[122] Vaccination was advocated by the newspaper.[123]

The number of cases was not extraordinarily high for such a large area – 100 by 4 August – but they were spread widely, and included a cluster of five policemen in Notting Hill.[124] Initially thought to be caused by cream, as the majority of cases had consumed the product from one original supplier, the cases continued, leading to doubts about this source.[125] A wide variety of foods and drinks were tested, including shellfish, fruit, vegetables and milk, cream and ice cream in addition to the water supply.

Criticisms included the recent exclusion of preservatives such as boric acid from milk and cream. A Ministry of Health order had been enforced to prohibit these substances but shops and houses did not have adequate cold storage to cope with this. Yet, these substances had been added to milk precisely because of the fears of bacteria in the late nineteenth century.[126] The epidemic drew a complaint from Kathleen Hilton Young of Lancaster Gate in west central London.

The Times published her letter which complained that her dustbins were only emptied once per week whereas in Westminster the dustbins were emptied daily. She claimed that it was 'surprising that the public health did not suffer', considering the recent hot weather.[127] It is impossible to know whether she thought the bins would cause noxious fumes in the heat or whether she was concerned about contact with the contents of the bins.

As was often the case, the cause of the epidemic was never confirmed. The most likely source was suggested by bacteriological tests and by epidemiology. The creamery had three suppliers in England, Ireland and Holland. The Irish cream was found to be suspicious, and included *B. coli*, indicating the presence of sewage.[128]

The epidemics studied so far have shown responses such as reluctance to pay for sewerage, dislike of chlorination and emerging litigation. The most enduring theme was sewer gas. Bacteriology was little use in proving the cause of epidemics until the carrier concept was defined, but even by 1927, it was difficult to absolutely confirm the cause of an epidemic.

Malton, 1932

The Malton epidemic was caused by excreta leaking from the Poor Law Institution into the water supply as a result of old and faulty drainage connections, which were probably shaken loose by heavy traffic. This contaminated the Lady Well with typhoid in 1932 as an inpatient in the Poor Law Institution was suffering from the disease in September and October 1932. Years before, an inquiry had noted the high consumption of water in Malton, so the drainage system had been faulty for a while.[129] Indeed, the well had been found to be polluted forty years before, but this had been forgotten and bacteriological tests had not shown problems since.[130] A newspaper report blamed the Ministry of Health for not taking note and enforcing change. There were also local discussions about asking the County Council for help as they had transferred the typhoid patient to the Poor Law Institution in the first place.[131]

The lack of confidence in the local water supply was expressed in letters to the local newspapers from early December 1932. For example, following the epidemic, a letter claiming to be representative of the Malton community complained about the complacency of the Malton Urban District Council towards the water supply and sanitation.[132] Another made the accusation that trade had been damaged because the problem with the water supply was not solved quickly.[133]

In early November, an emergency town council meeting was held at the town hall with the purpose of announcing the result of the bacteriological examination of the water supply. Despite the interest of the press in these results, the science was not as useful as epidemiology.[134] The emergency meeting was reported on as part of a full broadsheet spread in the *Malton Messenger*, but the water was shown to be in a normal condition.[135] Even though chlorination had been used for some water supplies since 1910, Malton's water had been chlorinated for the first time in response to the epidemic and in consultation with W. V. Shaw from the Ministry of Health. The reservoir had been emptied and the fresh supply of water had been chlorinated. The MOH, L.C. Walker, mentioned that men who served in the Great War would have experienced chlorinated water. The successful 'scientific method' of chlorination was developed during the war. Previously, chlorinated lime had been added to water which resulted in an 'objectionable taste'. Presumably some people were worried that chlorine gas had been used in gas warfare as he dismissed this concern and explained that the excess of chlorine was removed by adding another sulphite gas. This resulted in the elimination of typhoid and other bacilli and in water free from the 'objectionable taste'.[136]

Although the water was chlorinated after the discovery of the problem, the public still wanted another source to be found.[137] Despite Walker's arguments about the water, the local community objected to chlorination partly because of the taste of the water and because it was still the same 'poisonous filth that sent them to months of purgatory' that they were forced to drink.[138] Indeed, following the epidemic the council deliberated boring for a new water supply, despite arguments from the Ministry of Health and from a correspondent to the editor of the *Malton Messenger* insisting that chlorinated water was completely safe. In his letter, F. S. H. Ward claimed

> All experts agree that our present water treated as it is with chloride, is as safe from typhoid germs as any other water supply could be. Some people certainly complain that it is distasteful to the palate, but I am given to understand that there is as much chloride administered in one dose of medicine to a patient every two hours of the day as there is in 500 gallons of water – so no great harm should come to us in continuing our daily dose.

A deputation visited London in December 1932, consisting of the chairmen of the council and its finance committee and the clerk to the council. They arranged an appointment at the Ministry of Health in order to discuss the provision of a new water supply and the financial cost of the epidemic. Experts at the Ministry of Health, including the chief engineering inspector, were concerned that the proposed new site for a borehole could also result in polluted water, as it was too close to the town. The next topic for discussion was the cost of hospital treatment and inoculations, with the council representatives asking for permission to acquire a loan which would be paid over five years in order to spread the cost of the epidemic amongst the ratepayers, but entailing extra costs in the form of interest on the bank loan. This was sanctioned by the Ministry of Health.[139] However, a local hairdresser raised an objection at the public inquiry which was held to discuss the new water supply in the summer following the epidemic. He argued that the Ministry of Health and the council were responsible for the outbreak and yet the costs were falling on the ratepayers. Why had the Ministry of Health not provided a grant instead of sanctioning a loan?[140] There was logic in his argument as the Ministry of Health report had acknowledged that it was more at fault than the local authority as it had not passed on information about past inspections of the water supply.[141]

Although the people of Malton did not become as litigious as those in Maidstone in 1897, or Croydon in 1937, there were various complaints, many arising from the local hairdresser. At the beginning of the epidemic the hairdresser informed a reporter from the *Yorkshire Gazette* that he had telegraphed the Prime Minister to request free inoculations because 'local sanitation [was] abominable' and he thought that it was scandalous that a doctor asked for a guinea to inoculate his family. Walker responded at the Urban Town Council meeting that the hairdresser was ignorant and that, as MOH, he was duty-bound to inoculate anyone who demanded it.[142] At the end of December, the hairdresser wrote to the *Malton Messenger* demanding an inquiry.[143] At the beginning of January he wrote to the same newspaper in order to argue that the bill for nursing would be £10,000 and the loss in trade was £20,000.[144] A few weeks later, another member of the community wrote a letter to the press arguing that the local authority had been secretive and that they needed to write a report in order to answer many questions, for example the state of the water supply and sewers, the retention of a person with typhoid at the Poor Law Institution and the way in which he was nursed, and why private warnings regarding boiling water and milk were in circulation before an official notice was issued.[145] He considered that an inquiry would be a distraction from dealing with the epidemic, and that it would be an extra charge on the rates, and probably would not prove anything.[146]

Ideas of compensation were growing in Yorkshire, even though there was charitable help for those who were assessed as having economically suffered. One mother, of Stillington, York, tried to claim compensation from the fund as her daughter was a probationer nurse away at the hospital in Malton and apparently caught typhoid there, which additionally blocked a vein in her leg. National Insurance help had ceased as the maximum six months' worth of payments had been claimed. The mother wrote that the people in her local area had contributed to the Typhoid Relief Fund, knowing a local girl was involved. She also asked whether the Workmen's Compensation Act might cover her. The word 'compensation' was used frequently in correspondence by both parties, and a sum of money was finally given from the Relief Fund. Roland Elston, one of the administrators of the fund, replied that if she was a secondary case the nurse may well be entitled to Workmen's Compensation but not if she was a primary case. A later note said that she was not entitled to this form of compensation but that she could try filling in an application.[147] In another case, a man wrote to the committee asking for 'more liberal compensation' for a 'needy' woman. He was told: 'I am afraid the Fund does not attempt to compensate victims of Typhoid. All that it can do is minimize as far as possible the unavoidable privation and duffering [*sic*].'[148] Complaints were received from a group of seven people, who argued that the funds had not been distributed fairly. Three families were to receive payments for life, and one for ten years. These people were asking if the relief fund was based on the 'principle of "compensation"', as if so, they had lost young family members, and hence their future wages.[149] Notes about deaths and the fund also reveal that money was not given to the rich, and not to those who were poor if the dead relative did not contribute to the household income.[150]

Examples of what would be covered by the fund include the amount that a disabled or deceased family member contributed to the household income which was means-tested according to other household income, costs of travel and accommodation to visit sick relatives, employment of a servant to assist with childcare, and burial costs.[151] However, the arbitrary nature of the application of these rules is illustrated by two applications. One letter was written on behalf of a lady who had visited from Dublin and had contracted the disease. In response an offer of help was made, stating that

> it need not be shown that Mrs B. is in real financial difficulties. We want to help all those who have suffered from this terrible scourge. At the same time we don't want to pay where money is not really wanted.

Mrs B. was well off and declined the offer.[152] This could have been a public relations exercise as the woman was a visitor. In contrast, in the case of a poor man who used to live in the Malton Workhouse, there was debate about whether to help him with the means to set up a small business in case he was a 'Ne'r-do-well', as they did not want to 'throw good money away'.[153] Although the committee was not willing to help everyone, they returned some money to donors including the National Union of Teachers and the British Legion as it was not needed, and so their actions appear to be guided by principle rather than just distributing available funds to those who had suffered from the epidemic.[154]

There are several reasons as to why the people of Malton did not express anger towards the medical profession and the council as the people of Croydon would five years later. George Parkin, a local doctor, died in the epidemic, which generated a very charitable response from locals and the national medical community. A total of £724.15s.2d. was collected from the local community, and £955.9s.3d. from the medical profession towards the education of his children.[155] Perhaps this resulted in fewer criticisms of the local medical community. In any case Walker, the local MOH, acted very quickly. The day after the first case was notified, he arranged for the bellman to go around the town on the following day, warning people to boil both water and milk, and notices stating the same were distributed twice that day. On the same day he took a sample of the town water supply for bacteriological examination, and the sanitary inspector asked questions about milk supplies. Milk was easily ruled out due to so many retailers.[156] There are many other examples of how information was distributed quickly around the town.[157]

The medical community also acted quickly in inoculating the local population. Walker authorized free inoculation and so the administrators of the Relief Committee wrote to local doctors to ask whether they had inoculated people prior to this announcement and whether they were attempting to recoup fees for the service. There were four positive replies, including one from Noel Forsyth who had inoculated 102 people between 23 and 26 October.[158] As will be discussed in Chapter 6, there was considerable debate regarding the merits of inoculation at the time of an epidemic in 1937, so this response from the medical profession is different from that in Croydon.

There may also have been less serious complaints because the economic depression perhaps caused people to band together during the further crisis of the typhoid epidemic. Unemployment numbers rose from a maximum of 300 in the years between 1930 and 1933, to 540 during and following the epidemic due to loss in trade.[159] A major problem was that people were boycotting the town.[160] A circular was issued in the village of Langton urging people not to go to Malton, never to take meals there, to make sure any food from there was cooked and water was boiled, and that children should not go to Malton at all if it could be helped. There were no people at 'hirings', and excepting the shooting gallery and coconut shy, the market was empty, indicating that people did not want to buy food in Malton.[161] Even by the following May, Malton was still being avoided by people from neighbouring communities.[162] This probably resulted in local solidarity, hence the success in fundraising for the sufferers.

A sum of £8,700 was collected locally and nationally by the Typhoid Relief Fund Committee to help those in need, to replace clothing and bedding destroyed by disinfection, to assist with funeral expenses and to aid tradespeople who suffered financially from the boycott of the town.[163] Many other contributions were made by local volunteers, of which only some are listed here. The committee published handbills approved by the local medical officer, explaining how typhoid was and was not communicated, and published two bulletins daily reporting on patients who were placed in hospitals a distance away from Malton. They also helped the local council to question where patients had purchased their milk. People volunteered to work at the hospital, and to collect infected bedding, disregarding the risk. Thirty people donated blood and volunteers drove 300 people from hospitals to homes. Events were held such as a dance, an auction of boxing gloves, a navity play and carol singing, all in aid of the epidemic, and the Archbishop of York kicked off a charity football match. In the months following the epidemic, even Prince George came to visit and took tea made from Malton water.[164]

Despite considerable focus on the water supply, traditional responses to epidemics continued. A correspondent wrote to a newspaper in January 1933 that people in Malton had wanted to flee the town due to the problems with the water supply and the length of time it took to resolve it.[165] Another example of continuity is yet another letter about sewer gas. A correspondent wrote to the *Malton Messenger* that 'Only by making the future secure against the ravages of floods, the emission of sewer gas from completely broken up sewer systems, as well as contaminated water dangers, can we restore confidence'.[166]

Bournemouth, 1936

In Bournemouth in 1936, an epidemic of typhoid spread by milk contaminated due to a carrier was not blamed on the carrier, who was indirectly responsible for the deaths of 51 people amongst 718 cases.[167] He was the head of a large household, had been an MP for Dorset and was a 'prominent businessman'. The Ministry of Health representative, Shaw, said that he was completely unaware and very distressed at having caused the epidemic, and helped them in the investigations.[168] Effluent from his home had fed into a stream from which cows drank, probably leading to the epidemic.[169] There were only eleven cows in the farm which was located by the stream, but they produced milk for a dairy which received milk from thirty-seven farms, infecting the whole supply. The carrier was discovered when the LGB representative investigated how the milk became contaminated. When the stream was tested in September and early October, it was found to be negative for the presence of typhoid bacilli. However, in mid-October it was positive for the presence of the bacillus. In the further investigations the people living in the nearby large house were investigated, but were found to have had the vaccine, making the Widal test useless. However, their urine and faeces were tested for the bacillus, the carrier was found, and then measures of prevention to stop him causing a further epidemic began.[170]

The carrier's identity was not revealed at the time, he was taught how to cope with it, and a sewage disposal plant was built at his house. In later conversation it was found that even his daughter did not know that he was the source. In the 1940s, Mr Long of Frowds Dairy, the distributor which had spread the bacilli, took legal action against the carrier. However, the carrier held high office and was friends with Winston Churchill. He was protected from the media and after a year-long case, Long received no compensation.[171] Further criticism occurred as although the doctors worked hard, poor communication led to claims that possible damage to trade and tourism were reasons to conceal the epidemic.[172]

From Epidemiology to Bacteriology?

Although bacteriology was especially useful in identifying the carrier in Bourne-mouth, in the preface to the Ministry of Health report on the epidemic it was stated that the cause of the epidemic was only suspected, not proven.[173] William Coleman has argued that bacteriology was 'destined later to conquer the epide-miological field', but except for the Bournemouth epidemic, epidemiology seems to have been much more important than bacteriology in identifying the causes of the scenarios above.[174] Epidemiology was composed of many aspects such as case-tracing, meteorology, discovery of a resolution to the outbreak, of the mode of transmission and the original cause of the epidemic.[175] Epidemics studied by other historians also demonstrate the lack of utility of bacteriology in finding the origin of an outbreak.[176] For water analysis, Hamlin has noted that 'bacte-riology was remarkably uninfluential'.[177] As has been illustrated, it was a general problem that when the typhoid bacillus was sought in an epidemic it was almost never found in the suspected water, especially as there was a two-week incubation period for typhoid.[178] Indeed, the dominance of epidemiology over bacteriology, which Hardy discusses for 1890–1905 in England and Wales, appears to have continued well into the twentieth century, with the usual formula of investiga-tion of 'sanitary conditions, sewerage and drainage, the disposal of excreta and household refuse', milk runs and water supplies', 'meteorological factors', and any possible changes in the 'local sanitary and environmental features'.[179]

Traditional and sanitarian responses to disease continued throughout the late nineteenth and early twentieth centuries, despite the new science of bac-teriology: in particular, flight and discussions of sewer gas were almost always a theme, particularly when water was the cause. Even the medical profession dis-cussed sewer gas until at least 1905. In Gravesend and Malton, the public were reluctant to accept chlorination of the water supply. Yet, the public were thirsty for knowledge of the results of bacteriological tests of the water supply in Maid-stone and Malton, and possibly in Marlborough, and in the early 1930s they also sought inoculations to prevent typhoid.

Bacteriology could be essential for investigations. Indeed, Richard Thorne Thorne, an inspector for the Medical Department in England in 1895, com-mented that epidemiological evidence was by then subject to confirmation by bacteriology, when epidemiology had been sufficient before.[180] The carrier trans-mission of typhoid meant that bacteriology could be vital in stopping a further epidemic, for example in Eccles and Bournemouth.

This chapter has also revealed the development of a litigious response to epidemics. This was not a linear progression as the evidence of the precise cause was sometimes uncertain, and in Malton where it was determined to have been as a result of faulty sewerage leaking the emanations of a patient in the Poor

Law Institution, the public did not try to unite and claim compensation. The following chapter presents a detailed study of a unique lay response to a typhoid outbreak, resulting in the first successful community-wide compensation claim for an epidemic.[181]

6 THE 'TYPHOID TRIALS': COMPENSATION CULTURE IN 1930S CROYDON

The Croydon typhoid epidemic of 1937 claimed 43 lives from 297 cases. Despite debatable bacteriological evidence at the time of the subsequent Ministry of Health inquiry, the outbreak was determined to have been caused by a typhoid carrier working on a well which fed into a water supply consumed by 40,000 people in South Croydon.[1] This chapter examines how, through knowledge of the transmission of typhoid and contact tracing, lay people researched the cause of the disease. Subsequently, the epidemic resulted in a test case and 260 claims for compensation from the local authority.[2] Although infamous for debates on lack of communication within local government, the epidemic has not been studied in any detail before, and the extent of the influence of the local community has not been highlighted.[3] An examination of the press, the inquiry and court records reveals the extraordinary nature of the reaction to this epidemic, and the way in which bacteriological evidence was used by lawyers.

The lawyer appointed by the Ministry of Health to lead the inquiry, Harold Murphy, KC, was accompanied by two expert assessors: the civil engineer Harold Gourley and Sir Humphry Rolleston, the eminent Addenbrooke's physician and Cambridge Regius Professor of Physic, who had been physician-in-ordinary to George V. The inquiry began on 6 December 1937, when the epidemic was still very active with at least eighty patients remaining in hospital.[4] Following the inquiry, Murphy reported to the Minister of Health, highlighting on the first page the theoretical importance of bacteriology in determining the cause of the epidemic:

> The immediate cause of the outbreak was a portion of the public water supply becoming infected by the typhoid bacillus. The infected portion was that derived from a chalk well at Addington. How that well became infected is a question that cannot be answered with absolute certainty, but all the circumstances and probabilities point so strongly in one direction that I feel justified in coming to a definite conclusion on the subject. That conclusion is that the well was infected by the fact that at the end of September and during October, 1937, men, one of whom was an active carrier of typhoid, were working in the well and that during large parts of such period water from the well, unfiltered and unchlorinated, was being pumped to supply.[5]

The value of bacteriology was also highlighted when the breakdown in local government communications regarding the cleanliness of the well and water supply led to lawyers using laboratory test results in order to demonstrate responsibility for the epidemic.

The story of lay, legal and medical understanding of typhoid in 1930s Croydon has been constructed from the press, the inquiry minutes and report, Croydon Corporation records and the resulting test case for compensation.[6] The inquiry transcript runs to 1,010 pages and the transcription of the test case to 478 pages. These are extraordinary documents for examining public understanding of disease as many witnesses were questioned, including members of the local community, doctors and engineers. Fifteen large scrapbooks belonged to Sir Walter Monckton, the lawyer representing Croydon Corporation. Volume one for example, by far the largest, contains 197 pages of newspaper cuttings with the local and national press broadly represented on a daily basis. Two more large volumes of cuttings were compiled by Croydon Corporation. This study of the scrapbooks is supplemented by additional research within the pages of *The Times* and local newspapers. As with the outbreaks studied in previous chapters, local newspapers include letters from concerned members of the public providing evidence of their understanding of disease. In order to explain the actions of the medical profession, local council and local residents, the following topics will be explored: an outline of investigations into the epidemic and the actors involved, the role of bacteriology and immunology, the media, compensation culture, the continuity of traditional responses to epidemics such as fear and flight, and lay people's critique of medical expertise.

Typhoid in Croydon, 1937

On 29 October 1937, Richard Rimington was diagnosed with typhoid. He was one of the first people to contract typhoid and die as a result of the epidemic. As an immediate response to his son's diagnosis, Charles Rimington, an employee of the Bank of England, began conducting amateur epidemiology. He visited sufferers, whom he personally knew, asking for information, and pinpointed the cause of the epidemic to the water supply, eliminating other causes along the way.[7] Rimington reported his findings to Croydon's MOH, Oscar Holden:

> My son has just been taken to the Isolation Hospital suffering from typhoid. The maid from No. 66 in the same road has recently developed typhoid and a little girl from No. 64 is suspected of having the same disease. The milk supply in all these cases is not the same, shell fish and watercress have not been partaken of, the only common thing appears to be water. Some operations in connection with the water supply have recently been carried out in this road. I shall be greatly obliged if you will investigate the matter thoroughly and inform me of any precautions we can take to protect the other members of our families. My Doctor, George Lewin, has been informed by me of these other cases. I have also heard that there has been another in Pampisford Road, but know no details.[8]

Rimington also investigated whether the victims of the disease had been in contact with the public baths, butchers, bakery, fishmongers, greengrocers and grocers. On 30 October he went to see Holden and told him that he was at his disposal for a fortnight to assist with the investigations. On the following day, Rimington insisted that Holden and Charles Boast, the borough engineer (part of whose job it was to look after the town's water supply), should investigate some bad smells and a cesspool. Rimington wanted to know if the local water supply had been contaminated by roadworks.[9]

A meeting of forty local residents was organized by Rimington for the evening of 31 October. He also invited Holden and Boast.[10] At the meeting, Rimington advised Holden to inform medical practitioners of the outbreak. Ronald Moss suggested a general warning to the people of Croydon, especially with regard to boiling their water, and asked about bacteriological tests of the supply.[11] He had started boiling his water and milk as soon as he had heard of the outbreak on 30 October, as he was knowledgeable about typhoid transmission, having lived in India for ten years.[12] Another local resident had also started to boil water and milk for residents at her nursing home on 29 October after she had discovered why Richard Rimington had been taken away in an ambulance.[13] Cecil Green suggested extra chlorination of water in the local area.[14]

The residents' further suggestions for causes of the epidemic were flies, food, leaky drains and a cesspool.[15] These citizens were bold in discussing their beliefs regarding disease transmission, even in the company of the MOH and the borough engineer. Indeed, Rimington recalled at the inquiry that '[o]ne man was very keen on flies'. Flies were also cited as a vector when Rimington informed Holden of the number of flies in the Croydon Borough Hospital where his son was admitted, especially as the hospital was situated next to the sewage farm.[16]

At the meeting and the inquiry there was much discussion about the bacteriological testing of the water. Moss asked Holden whether he thought that water was the cause. According to Rimington, Holden answered that it was 'inconceivable' that water was the means of transmission as the water was tested once a month.[17] Having spent so much time in India, Moss was 'amazed' at this answer and asked whether 'any special bacteriological examination had taken place since the outbreak of typhoid?' Holden referred this question to Boast who answered that the water supply was routinely tested.[18] Moss was not satisfied with this response as he connected 'routine' with meaning 'slipshod and not necessarily a very detailed examination'. So he asked Boast 'whether some bacteriological examination ought not to be made ... before this meeting on the 31st'.[19] The Outbreak Committee put the same question to Alderman Wood Roberts, the chairman of the Water Committee, when they visited as a deputation. They were told the water supply was in fact tested on 29 October and that although *Bacillus coli* had been found, the test was not conclusive for typhoid.[20] At the inquiry Moss also remembered that Roberts told them that the tests had occurred every fortnight and that this was 'abandoned on grounds of economy'.[21] At the inquiry, Holden's representative, lawyer Geoffrey Howard, complained that Moss made an inaccurate claim about these events in the Outbreak Committee's Memorandum for the Ministry of Health. Howard argued that Boast had answered this question about the bacteriological investigation and Holden had already ordered the test of the water supply in St Augustine's Avenue, Rimington's address, proving that he was taking this possibility seriously.[22] Monckton, on behalf of the Corporation, argued that the Water Committee minutes showed that the samples had never been taken fortnightly, so this practice certainly had not been abandoned.[23] However, despite the confusion in remembering these events, it was the case that Holden was unaware that the water from the well was not being chlorinated at that time, whilst work was carried out.[24] The Ministry of Health report concluded that Holden, the MOH, was ignorant of the fact that water is the most common cause of transmission; the recent typhoid epidemic in Bournemouth in 1936 had led Holden to believe that milk was the most common cause.[25]

With regard to the community's ideas about the transmission of typhoid in the nineteenth century, 'flies, fingers, and food' were demonstrated to be vehicles that spread pathogenic bacteria, and these vehicles gradually replaced the miasmatic ideas that filth and sewer gas spread disease.[26] Yet the discussion of leaky drains and a cesspool still indicate beliefs in miasma in the 1930s, though it could also indicate concerns about the contamination of groundwater and the water supply. Other methods of protection from disease, such as boiling water within the home, and to some extent awareness of the dangers of milk, preceded bacteriology, as discussed in Chapter 6.[27] People could have been concerned about the role of the fly prior to 1937 but concerns about this means of transmission were not recorded in any of the sources examined for Bangor, Marlborough College and Malton. Therefore, whereas many of the methods Croydon citizens raised to protect themselves from disease could be thought of in terms of everyday hygiene rather than influenced by bacteriology, the discussion about flies in particular suggests lay engagement with the discoveries of bacteriologists.

Flies became 'scientific and dangerous' during the late nineteenth and early twentieth centuries. Not only was the housefly confirmed by scientists to be spreading disease, but also other insects such as mosquitos, rat fleas and lice.[28] In 1902 Alice Hamilton studied the role which flies played in a typhoid epidemic in Chicago. She captured flies and drowned them in eighteen test tubes, isolating typhoid bacilli in five of them.[29] In 1909, J. Nash cultivated 'myriads of germs' including intestinal bacteria from a fly found in a hospital ward.[30] By 1914, five LGB reports in Britain had investigated the habits of the fly and the bacteria which could be cultivated from them.[31] The pathologist Graham-Smith, whom we encountered in Chapter 1, wrote *Flies in Relation to Disease: Non-bloodsucking Flies* in 1913. In the same year the British Museum published a book on the dangers of this insect and erected a large model of a fly accompanied by a tray of food and kitchen waste in the Central Hall, and in 1915 an anti-fly exhibition was hosted by the Gardens of the Zoological Society. From 1915 to 1917 articles regarding the dangers of flies appeared in *The Times*.[32] In particular, a campaign ensued in the press when flies were connected with disease in trench warfare.[33] Hence there was a transition in beliefs about flies from their portrayal as 'God's bin men, designed to remove putrefying matter' and to ventilate buildings with their buzzing wings, to the realization that they were a threat to health.[34] Major G. Hurlstone argued in *Book of the Fly* (1915) that readers needed to forget the religious idea that everything had a purpose.[35] Persuading the public of the danger of flies was not easy; an American entomologist argued that the housefly needed to be renamed the 'typhoid fly' in order for the public take the matter seriously.[36] Yet, as with many other ideas of disease transmission, concepts of the fly as a carrier of a disease had been indicated prior to bacteriological evidence, when in the 1840s Henry Mayhew connected flies with filth and poverty in his *London Labour and the London Poor*, arguing that grubs grew in excrement.[37]

To recapitulate the events of the last three days of October, Rimington, rather than the MOH, informed the borough engineer about the typhoid outbreak and the two local government experts met at this point because of Rimington's interventions. Boast only discovered the problem with the water supply due to Rimington's suggestions that it should be investigated.

On 1 November, Boast communicated with his staff, discovering that chlorination of the water supply from the Addington Well had ceased. Following a discussion with Holden, chlorination was resumed on that day. After there were two more notifications on the same day, Holden informed the Ministry of Health about the outbreak. He also wrote letters to all the registered medical practitioners in the borough of Croydon, advising them to bear typhoid in mind.[38] In his Ministry of Health report, Murphy criticized the letter, stating that many practitioners would not have seen a case of typhoid before and that they were not warned about the risk from water. He also pointed out that some residents' doctors may have worked in the surrounding boroughs and therefore would not have received information about the outbreak.[39]

After another notification on 2 November, Holden requested assistance from one of the Ministry of Health's medical staff. Ernest T. Conybeare, the Ministry's expert on typhoid, visited Croydon on the following day and continued to visit throughout the epidemic. He immediately considered water as the cause and asked for a map of Croydon's supplies.[40] On 3 November, it became clear that the high level supply (rather than the low level supply) of water to Croydon was a common element amongst the typhoid cases. Conybeare initiated the collection of samples from two wells, which were sent to the Ministry of Health's bacteriologist. On 4 November, the tests revealed that while one well was satisfactory, the Addington Well was 'heavily polluted', although the typhoid bacillus had not been isolated. The supply from the well was cut off on that day and there were thought to be no more primary cases. In any case, work had finished on the well and the polluted water supply had been chlorinated since November.[41]

Typhoid bacilli were sought in various locations near to the well. Workmen for the education department had been using a latrine almost directly above a well adit (a tunnel to direct water from water-bearing fissures into the well), and privies at a farm drained into soil not far from the well. More than 200 samples were taken in the area of the well, plus more from storage tanks and taps filled by Croydon's high water supply. All of the samples – including testing of all the occupants of a gypsy camp, and the tracing of the scattered workmen who had been working at St Giles' School – were negative for the typhoid bacilli.[42]

Murphy and his assessors were concerned that water remaining in the system, for example in pipes and cisterns, could cause further cases of typhoid. Most houses had cistern storage capacities of up to 26 gallons. On his arrival on 11 November, one of the first pieces of advice from Ernest Suckling, the Ministry of Health's expert on water and bacteriology, was that water from temporarily unoccupied houses should be run off before use as a precaution, with a notice to that effect in the press. Perhaps as a result of this action, the inquiry did not reveal any cases within any of the thirteen unoccupied houses with storage tanks.[43]

Holden and Conybeare were not informed about the work that had been carried out on the Addington Well until Friday 12 November. Therefore the limited communication between Boast and Holden continued. The weekend caused further delay, and on Monday 15 November, Conybeare asked for the workers' names and addresses and arranged to test their blood. These eighteen blood samples were sent to William MacDonald Scott, the bacteriologist for the Ministry of Health. On 16 November, Scott wrote to Conybeare as there were eight Widal's tests with reactions 'of interest', of which four were selected for further investigation. These men were admitted to the Mayday Hospital for further tests on their faeces and urine samples which were collected and sent to Scott. Bacillus typhosus was found in all five of the faecal samples from one workman, but in none of his twelve urine samples.[44]

In the meantime, a group of lay citizens in Croydon formed the South Croydon Typhoid Outbreak Committee within a few days of the residents' meeting. Rimington was appointed as the chairman. The committee met every night for nearly a month. They wanted publicity for the outbreak and 'placarded the whole town'.[45] On 12 November, Rimington led a deputation to the alderman responsible for the Croydon Water Committee. The Outbreak Committee wrote a letter to the Ministry of Health on 17 November to justify their petition for the inquiry, and on 18 November, Rimington led a deputation to the Ministry.[46] Additionally, Rimington went to see his doctor, Lewin, twice a day to inform him about new cases. Lewin announced at the inquiry that Rimington gave him more information about the epidemic than the MOH.[47]

Rimington and others were particularly annoyed about the lack of precautions disseminated to the people of Croydon.[48] Although they were knowledgeable about what to do, they thought that others would not be. In his original letter to Holden, Rimington had asked for instructions, and never received a reply.[49] A statement of precautions was finally released on 5 December, with advice about hand-washing.[50] Montagu Lyons, the lawyer representing the Outbreak Committee, told the Chief Sanitary Inspector, Robert Jackson, 'The one thing that stands out a mile as a possibility in the case of any typhoid outbreak, you thought fit not to give a single warning or instruction', including little effort to warn the schools.[51] However, Jackson revealed that there was one place which had received precautions instructing the boiling of water: the residential Russell School which was situated right next to the contaminated Addington Well.[52] Although Richard Rimington's school, the Whitgift School, had received no instructions, the school announced on 20 November that all water at dinner was being boiled, and in any case the 'ordinary' water supply had been chlorinated from 4 November.[53]

The topic of the outbreak was raised in Parliament for the first time on 18 November by Harry Day, Labour Party MP for Southwark, who was assured that an inquiry had already been ordered.[54] More questions regarding collaboration between private doctors and the public health system, and requests for public representation at the inquiry, were posed on 25 November.[55] On 30 November, Day asked whether the Minister of Health was aware that the inhabitants of Croydon had not been immediately instructed to boil their water after the epidemic broke out and was again told to wait until the inquiry.[56] Further enquiries were made regarding the provision of a specific water engineer in other local authorities, the case of a boy from Jarrow who caught typhoid and died whilst in a Croydon hospital following a road traffic accident, a question regarding whether the Addington Well water had been reintroduced into the Croydon water supply, and a proposal that local authorities needed a standard of purity for water.[57]

The Significance of Bacteriology

In the Croydon case, epidemiological methods were used both by lay people and the Ministry of Health's typhoid expert in order to pinpoint the cause of the epidemic to the high water supply. Almost exclusively, the only people who became ill received this water in their homes or had occasionally drunk from the supply. Epidemiological methods of case tracing and assessment of environmental conditions still seem to have been much more useful than bacteriology, at least until phage typing for typhoid was developed in 1938. Despite the utility of bacteriology for diagnosis, identifying potential carriers and suggesting problems with water, milk and food, when the typhoid bacillus was sought during epidemics it was very rarely found in suspected water. The disease has a two-week incubation period and the bacillus only survives for around five days in water.[58] So was bacteriology useful in resolving epidemics?

On 4 November, water from the Addington Well had been tested by Scott, and was found to be heavily polluted with *Bacillus coli*, a more resilient bacteria than typhoid which was an indicator of contamination by sewage.[59] A letter from Suckling was read out at the beginning of the inquiry, stating 'I am satisfied that since November 11 1937 the public water supply at Croydon has been bacteriologically pure and wholesome and suitable for drinking and domestic purposes'.[60] This was the date when Suckling came to Croydon to help.

At the inquiry there was an expectation that bacteriology would provide the evidence of how the epidemic began. This belief was also disseminated in the national press in November when the *Sunday Times* reported that the town clerk claimed that tests for the 'mystery carrier' were being carried out 'night and day'.[61] The laboratory was used in order to identify a particular carrier, although the man's role in causing the epidemic was not absolutely proven. There was much discussion about bacteriological tests at the inquiry as repeated tests only showed the bacilli in the carrier's faeces, not his urine. The workers were only allowed to urinate, not defecate, into a bucket in the well, and the workers, including the carrier, claimed to have strictly followed these instructions.[62] A man had died in January 1937 when he was lifted up the well shaft, so it is understandable that the workers would not want to make extra trips, as was pointed out in the test case. A report from Sidney Rafferty, the water engineer, was read out, in which Rafferty argued that with spells of 3.5 to 4 hours down the well, and 7 to 10 minutes needed in order to be lifted out of the well, the sanitary arrangements might not have been effective.[63] Humphrey Brandram-Jones, the chief water assistant during the early work on the well, revealed that if a man needed to urinate, the bucket had to be sent for and then returned to the surface. However, there was no provision for the bucket to be covered whilst the urine was lifted out and it could have tipped or slopped urine into the well. Furthermore, the carrier could have got faeces on his clothes or hands after using the lavatory.[64] The lawyer, H. J. Wallington, argued with Brandram-Jones that he could have disinfected the bucket to make it harmless if the urine was spilt, and suggested that it was common knowledge that human waste could be dangerous to health: 'you are a human being and you understand that there is a possibility of pollution of water if material of this kind gets into it', to which he answered 'Quite'. Brandram-Jones protested that he could have had 'special containers' for the purpose, but that he had nothing to guide him, and that other engineers were not practising any different methods.[65]

William MacDonald Scott, the bacteriologist for the Ministry of Health, testified on 7 January 1938. As substantial evidence had already been provided by Scott's colleagues with regard to bacteriological test results, he was only asked to comment on the complicated topic of strains. Scott had written a report to the Ministry of Health on 24 November detailing his tests using twelve swabs from six patients, which he compared with the organisms isolated from workman 'A', the suspected carrier. He found that the fermentation tests on all of these samples gave identical results. Murphy asked Scott to comment regarding the significance of this result. Scott explained that if there had been differences this would have meant that there were different origins of the illness, but that identical results did 'not mean very much, because the majority of typhoid bacilli taken from all over the country would give characters exactly like that. It does not help very much.' Adding uncertainty to the identity of the carrier, there was an occasion when one of the other workmen had a positive result for typhoid bacilli in his faeces which was also the same strain.[66]

A typhoid strain with the Vi antigen (Vi standing for virulence because it was much more resistant to antisera) was discovered by Arthur Felix and Margaret Pitt at the Lister Institute in Britain in 1934.[67] However, although this distinction was used by the Ministry of Health pathologist to confirm the link between a carrier and the epidemic in the 1936 Bournemouth epidemic, this was not discussed with reference to Croydon.[68] After phage typing for typhoid was developed during the year after the Croydon epidemic, the first practical application of the technique occurred in January 1939 when a laboratory worker believed he had contracted typhoid at work. However, it was discovered that the laboratory only held types A, E and F whereas the patient's strain was C, and therefore this was not contracted from the laboratory.[69] Thus, these techniques could be useful if the perceived cause was found to have a different strain from the outbreak.

In order to clarify the decisions made about bacteria at the inquiry, John William Henry Eyre, Emeritus Professor of Bacteriology at the University of London, was called as a witness for the test case. Eyre was involved in the investigations into the typhoid epidemics in Maidstone and Lincoln in 1897 and 1905. He was asked to describe the transmission of typhoid, including the creation and role of carriers, and the course of the disease. He also explained the agglutination blood test for typhoid sufferers and the transmission of typhoid in water in laymen's terms.[70]

Immunology was also discussed in the inquiry and in the lay and medical press. Although the typhoid vaccine had been available since 1898 and was used widely in other epidemics during the 1930s, the Ministry of Health's advice during the Croydon epidemic was not to inoculate. This topic was discussed in the House of Commons. Wilfred Roberts, MP asked the Minister of Health whether, in light of the success of inoculation in the Basque camp at Eastleigh in the summer, inoculation would be recommended in Croydon. This camp was for children who were refugees from Guernica, a town destroyed by bombing during the Spanish Civil War. The Minister of Health, Sir Kingsley Wood, issued a written response:

> The circumstances of the two outbreaks are different, and I am advised that general anti-typhoid inoculation in Croydon would not be beneficial. The protection by inoculation or otherwise of individuals resident in houses in which cases of typhoid are being nursed is a matter for consideration by the medical practitioner in attendance, in the light of the special circumstances of each case.[71]

The use of typhoid inoculation was also decided against during a simultaneous typhoid epidemic in Kensington, London.[72] On 21 November, the *Observer* published a statement from Holden arguing that after consulting with 'several medical colleagues ... the weight of opinion is against its widespread application' because of the negative phase.[73] At the inquiry, Holden declared that the negative phase – the idea that people could be more susceptible to typhoid just after the inoculation – was 'a well-known medical fact'. The topic was discussed very briefly, with Lyons pointing out that there was actually an ongoing debate regarding the idea of the negative phase. Holden conceded that this was the case.[74] When the same statement was published in *The Times* on 22 November as that which had been published in the *Observer* on the previous day, an immediate response was provoked.[75] A. D. Gardner of the Standards Laboratory of the Medical Research Council at Oxford University wrote to *The Times* the next day. Published on 25 November in a letter to the editor, he argued against the publicity in the newspaper from medical men who believed in the negative phase, arguing that there was no proof of this. He mentioned that there seemed to be a fear of this inoculation in the face of the epidemic, and argued that the people under threat should be inoculated.[76] Following the publication of an article in the *British Medical Journal*, which argued that inoculations were usually given only when epidemics were under control and that they could aggravate an attack of typhoid, a series of letters advocating inoculation appeared in the journal.[77] S. Watson Smith of Bournemouth wrote to suggest that those in the profession 'old to the ways of typhoid' knew that there was no 'terror' regarding the negative phase and E. W. Goodall described the idea as a 'bogy'.[78] During the Bournemouth epidemic, most of the nurses and maids at the Royal Victoria and West Hants Hospital had been inoculated and none contracted the disease.[79] Likewise, Richard Taylor discussed inoculation policy at the Basque refugee camp. The Ministry of Health advised inoculation of people working in sanitation and with food, when two days after the children's arrival cases of typhoid were diagnosed. The Ministry did not advise inoculation of the 3,280 children in the camp. However, local committees which had agreed to look after the children insisted they were inoculated before leaving the camp. Therefore Taylor arranged for general inoculation and there were no further cases.[80] Major Greenwood, the leading medical statistician in the country, contributed to the controversy in December 1937, clarifying the analysis of statistics on typhoid fever contacts and inoculation, perceiving that the number who contracted typhoid was indeed significantly less than the expected rate had they not been inoculated.[81] The *British Medical Journal* summarized that there was no evidence of a severe attack of typhoid following inoculation and that official policy on this topic needed to be declared.[82] Following the epidemic, the debate regarding the negative phase continued. William Wilcox, a consulting physician to

St Mary's Hospital and senior physician at the London Fever Hospital, argued against inoculation during a potential incubation period. He had treated sixteen of the Croydon cases, as these patients had been admitted to the London Fever Hospital. Wilcox had seen patients who suffered with typhoid following inoculation in the South African War and these cases had been severe.[83] He also gave evidence at the inquiry, and argued that typhoid inoculation did not completely protect against the disease anyway, but could lead to milder cases.[84] After the inquiry, the topic was raised again in Parliament. Thomas Groves, MP asked Wood how many people who were inoculated with either anti-typhoid vaccine or anti-typhoid serum developed typhoid during the epidemic. However, the result of Wood's enquiry to the local authority was reported directly to the member and was not included in Hansard.[85] Despite this debate, according to the *Sunday Times*, 'scores' of people were getting vaccinated every day in 'desperate hope' of protection against the 'peril'.[86]

Other immunological products were discussed. At the inquiry, a man called Mr Harris was given permission to ask Holden a question regarding his adoption of bacteriophage inoculation which had recently been used in India, Brazil and Pennsylvania. Holden was not using this inoculation.[87] J. V. Pincus, from Clapham wrote to the *British Medical Journal* about an oral version of the TAB vaccine which had been developed by Alexandre Besredka twenty years earlier. It had not been used in Britain because of the low incidence of typhoid, but was used in tropical and subtropical countries. Pincus had spent time working in South Africa where oral vaccines were given to doctors for free in order for these to be distributed widely. He argued that these vaccines resulted in less side effects, less pain, were quicker to administrate, were effective within five days and that people were less resistant to this form of the vaccine in comparison to the subcutaneous form. The editor of the *BMJ* added a comment to Pincus's letter reminding readers that an article on 10 April 1937 had argued that the oral form and an ointment form were probably useful.[88] Ernst M. Fraenkel also proposed that the method had been successfully used in France, Germany, Eastern Europe and Japan.[89]

The utility of typhoid serum has been debated since at least 1909. Doctors found very uncertain outcomes.[90] The serum was used as a form of inoculation in the Kensington outbreak and no one who received it became ill. However, the doctors involved acknowledged that this was not proof of a preventative measure as the thirty-one people may not have developed typhoid in any case.[91] Yet 'a special anti-typhoid' serum had been obtained from the Lister Institute 'some days ago' and was distributed to local doctors, but it was too early to tell whether it was helpful.[92]

One doctor who corresponded with the *British Medical Journal* claimed to have successfully used sulphanilamide in order to treat typhoid.[93] Wilcox argued that the drug was ineffective against typhoid in humans, and possibly harmful to the haemoglobin, leucocytes and the heart muscle, even though tests with mice showed potential.[94] However, he was opposed to any interference with typhoid by using drugs or vaccines, indicating his non-interventionist approach; when considering use of the serum, he was concerned that it might alter the 'delicate mechanism of immunity'.[95]

The inquiry also resulted in a suggestion that bacteriology could be of use in the future in order to prevent a reoccurrence of typhoid. Murphy argued that workers should receive a Widal's test to see if they were carriers of the disease and, of course, proper disinfection of the water supply should take place as usual if work was being carried out.[96] Although the American case was not mentioned in comparison, this directly follows the course of action that New York City took in response to Mary Mallon's recapture after she caused another epidemic in 1915. By the end of the following year, the city had tested the urine, faeces or blood of 90,000 food handlers. Despite Mallon's incarceration after her second apprehension, as in Croydon, the city was held to blame. Indeed, the *American Journal of Public Health* compared ineffective means of control in the USA with police systems in Germany and Europe.[97] In the case of Croydon, the local government accepted all responsibility, with the carrier exonerated from responsibility, and subsequently the town followed the same idea of control for those working with water as New York instituted for food handlers. Laboratory tests were the key to future prevention in both cases.[98] Indeed, coping with carriers meant that bacteriology could be vital in stopping a further epidemic, as the Bournemouth epidemic showed in Chapter 5.[99]

Typhoid and the Media: 'The Death Bulletin'[100]

It was not only the medical press that was filled with debate about typhoid. As has already been mentioned, local and national newspaper cuttings filled seventeen scrapbooks. In early December, the public press was so full of stories about the epidemic that in the House of Lords, the Lord Bishop of Winchester used the epidemic to argue about the lack of press on road traffic accidents, for which there were many more casualties.[101]

Initially, Holden and the Public Health Committee did not want to cause 'panic' by alerting the media. They suffered criticism for this strategy.[102] The epidemic was not reported in the local *Croydon Advertiser* until 6 November, when the report made the front page. There were ten cases by that date, two of whom were seriously ill. The residents were told that the situation was 'well in hand'. Holden told the reporter that the cause had not been discovered but that it was out of the question that it was food or milk and, despite rumours, typhoid had not been introduced into Croydon by a maid who had recently arrived from Holland. Holden and Alderman Bessie Roberts, chairman of the Public Health Committee, were disappointed by the 'alarmist rumours' which were already causing a scare. Roberts said that there was 'No concealment of facts ... by the authorities, and statements which have been circulating that the outbreak is much more widespread are definitely untrue'.[103]

Holden decided to delay announcing the outbreak until a cause had been established. According to Rimington, at the meeting of local residents Holden rejected the suggestion of telling people to boil water as it would cause panic.[104] Although problems with the water supply were discovered at the beginning of November, the Public Health Committee agreed not to tell people to boil their water, due to chlorination making it unnecessary. Holden argued that the water was chlorinated on 1 November, before it was known to be unsafe.[105] As the South Croydon Outbreak Committee's lawyer, Montagu Lyons, argued, there was still water in the system from before chlorination.[106] Earlier samples were not taken from water supplies at affected houses, again to avoid 'anxiety' as people may have thought their water was not 'in a proper condition'.[107] Even one of the town councillors only found out about the epidemic via the press on 5 November, despite attending all council sessions. He was only informed through official routes on 9 November.[108]

This policy of concealment was not infrequent in epidemics in the bacteriological age, such as those of cholera in Hamburg and Naples. There were various reasons for concealment, including avoiding panic and damage to trade.[109] Jacob Gould Schurman, president of Cornell University, tried, unsuccessfully, to conceal the 1903 typhoid epidemic from the nation, but not from the people of Ithaca and parents of students at Cornell University. There were also accusations from Ithaca's citizens that university experts had covered up the signs of the epidemic as the university had investments in the private water company; claims which were denied.[110] Even if concealment was not deliberate, it appears to have been easily suspected in epidemics, perhaps showing the general public's view of how local and national governments would act in a time of epidemic crisis. This has been seen in the analysis of the Bournemouth epidemic in Chapter 5. The wish to avoid panic in Croydon must have led to a larger number of cases than necessary.

As feared, the media contributed to panic in Croydon and elsewhere when it learnt the news of the epidemic. National newspapers, especially those which published photographs such as the *Daily Sketch*, portrayed Croydon as a dangerous and frightening place to be (see Figures I.1 on p. 1 and 6.1 below). Figure 6.1 graphically illustrates the inhabitants' fears. They are shown anxiously reading the typhoid notices outside the town hall. Everyday passers-by stopped to scan the board, hoping to read that the epidemic was over. The *Daily Express* referred to notices on this board as the 'Death Bulletin'.[111] In Figure I.1, the town clerk (the modern-day equivalent is Chief Executive), E. Taberner, is pictured actively engaged on the phone in dealing with the epidemic, and a laboratory worker can be seen testing the water supply.

Figure 6.1: 'Croydon's Typhoid War', *Daily Sketch,* 27 November 1937. Source: *Daily Sketch*, 23 November 1937, Sir Walter Monckton, KC, 'Outbreak of Typhoid in Croydon, Nov. 1937, Press Cuttings Nov. 4–25, Local Papers Nov. 26–27, Volume 1, Town Clerk Croydon', Croydon Local Studies Library and Archives Service, fs70 (614.4) CRO, p. 113. Image reproduced courtesy of Croydon Local Studies Library and Archives Service.

A series of articles created panic about watercress and reported the downturn in sales of the food in the *Morning Advertiser, Star, Sunday Chronicle, Sunday Pictorial, Reynolds News, Sunday Express* and the *Manchester Guardian* on 20 to 22 November.[112] These articles followed an announcement in the BBC News, after a statement by Taberner that 'Three cases of typhoid are reported in one family living outside Croydon. It is believed that each has eaten watercress stated to have come from near Croydon.'[113] The *Daily Express* announced that '[i]solated cases have caused fear in the City of London'. For example, a Croydon girl had been taken ill while working at Lloyds in the City. Inciting panic in the capital city, the article began by highlighting that Croydon was only 'ten miles south of London Bridge'.[114]

Two letters to the editor of the *Croydon Advertiser* aimed to stimulate retribution. Freda Churchill wrote in to say that residents were told that the epidemic was 'well in hand', and that there was a statement in a daily paper that there was no need to boil water. She believed that the Health Department was to blame for residents not starting to boil their water earlier:

Had this been done the outbreak might not have assumed the proportion which it has done – with fatal results in some cases. One wonders if the confidence which is reposed in the Health Department is justified by these recent events.[115]

There were many more letters about the 'lack of quick thinking', and an editorial commented that public announcements about precautions from a loudspeaker van should have been used even though it would have caused panic.[116] Rimington wrote to thank readers for their sympathetic letters about Richard's death, and asked the residents of Croydon to sign the petition for a public inquiry which was available in the offices of the local newspapers and various shops in Croydon.[117]

The epidemic was also used for political propaganda. The *Daily Worker*, a socialist newspaper, adopted the epidemic for its own cause, inciting further panic. It claimed all London was at risk, argued for more publicity, and blamed the local Conservative-led council and 'Tory' economic cuts for causing the outbreak, repeating the claim that the frequency of inspections of the well were reduced due to the cost.[118] By March 1938, the local Conservative Party members were referred to as the 'Typhoid Tories' in a headline. They apparently feared their own supporters: 'the ultra-respectable Croydon Typhoid Outbreak Committee'.[119] The Labour Party used similar tactics, citing the epidemic in propaganda for a bi-election campaign in December.[120]

The Chamber of Commerce for Croydon, whose members were losing money through depleted trade, campaigned against sensational newspaper reports. A news agency was recruited to help on 24 November, and the Chamber went ahead with its annual banquet amidst much publicity. The BBC was even used to broadcast its viewpoint.[121] The Ministry of Health also used the media, releasing a statement to the press that 'alarmist suggestions' were damaging trade. The Ministry of Health, in Murphy's words, wanted to 'dispel the delusion which does still continue to exist in the minds of some people – I hope a minority – that typhoid is something that one can get ... by mixing in shops and other public places'.[122] In the House of Commons, Patrick Donner claimed that the typhoid epidemic had brought the whole watercress industry to a standstill and that he was approaching the British Broadcasting Association and the public press in order to disseminate the idea that watercress grown outside the Croydon area, which was clearly labelled with the place of origin, was not dangerous. Major-General Sir Alfred Knox also raised concerns about the watercress industry and asked for a public statement via the BBC.[123] The fruits of this campaign can be seen in Figure 6.2. Exactly the same photograph, illustrating the local Chamber of Commerce banquet, appeared in at least four national newspapers on 25 November: the *Daily Express*, the *Daily Sketch*, the *Daily Mirror* and *News Chronicle*.[124] Similar concerns were raised about the oyster industry which also suffered due to widespread fear of typhoid, despite there having been no suggestion that this epidemic had been caused by oysters.[125] Other articles which were clearly part of this campaign included 'Strong Protests Against Alarmist Typhoid Statements: Section of the National Press Blamed' in the *Croydon Times*,[126] and in the national press, 'Shop, Play, Travel as Usual' in the *Daily Mail*,[127] all published on 24 and 25 November. One shop, Kennards, offered insurance against typhoid for shoppers spending a certain amount. They quoted a letter written by Thomas Horder to the Chamber of Commerce, where he expressed his concerns about erroneous beliefs that typhoid could simply be caught by going shopping.[128]

Figure 6.2: 'They Ate Cress at Croydon Banquet', *Daily Express*, 25 November 1937. Source: *Daily Express*, 25 November 1937, Sir Walter Monckton, KC, 'Outbreak of Typhoid in Croydon, Nov. 1937, Press Cuttings Nov. 4–25, Local Papers Nov. 26–27, Volume 1, Town Clerk Croydon', Croydon Local Studies Library and Archive Service, fs70 (614.4) CRO, p. 161. Image reproduced courtesy of Croydon Local Studies Library and Archives Service.

The local press presented a very different picture to the national newspapers, arguing that the people were not being swayed by sensationalist reports. The *Croydon Times* reported on 27 November that the people 'refused to be misled by sensational statements which have been published. In this respect, Croydon people are proving themselves to be a commonsense community'.[129] However, as will be discussed, people were very frightened by the epidemic, avoiding contact with others if possible. In response, the newspaper may have tried to allay this panic by praising the conduct of Croydon residents.

The *Croydon Advertiser* disseminated bacteriological knowledge before the end of November. Edward Burnet, a Harley Street physician and 'one of the leading pathologists of the country', wrote an article for the paper which was published on 26 November. Burnet had helped the Home Office with the Limerick typhoid outbreak inquiry before the war. He believed that the more people knew about bacteriology and the transmission of typhoid, the less they would panic. He gave details of how to prevent typhoid which were not significantly different from those which the public discussed at the citizens' meeting. Advice included a discussion of water, milk and dairy products, oysters and shellfish, and the 'feet of flies'. He argued that if cleanliness was observed, typhoid sufferers could be cared for with no risk. Chlorination of water, boiling of water and milk, avoiding salad and raw vegetables, and hand-washing were also suggested as means of prevention. The physician bemoaned the hysteria which had led to problems for shops and leisure venues, and the article is accompanied by a notice from E. Taberner, the town clerk, reiterating Burnet's instructions and the confidence the public could have in the shops and amusements in Croydon.[130] Burnet's article echoed a letter written by Sir Kaye Le Fleming, the chairman of the British Medical Association Council, and Lord Bernard Dawson, chairman of the Royal College of Physicians. In 1920, Dawson had published a report which had proposed clinics within a system of local primary, secondary and tertiary health care. In his letters to *The Times* he argued that the Croydon epidemic highlighted how the findings of his report had been neglected.[131] Burnet revealed similar sentiments regarding the lack of communication between local doctors which could sometimes let down the Public Health Service.[132] Nevertheless, he tried to restore confidence in the medical profession with his article and advocated positive thinking:

> Interference with the normal conduct of life, with your relations with your tradespeople, or with the pursuit of your usual recreations, is absolutely unwarranted. If you are ever in doubt about yourself ask your doctor, and he will inspire you with the confidence which is no mean factor in immunity against disease.[133]

Nationally, the *Star* ran an article on 'How Medical Detectives Work' which explained the concept of typhoid carriers, bacteriological blood tests and water supply tests.[134] Residents also wrote letters to the local press with ideas of how to allay the epidemic. Philip Dawson, a member of the National Association of Medical Herbalists, suggested that there were ways of making typhoid germs unable to live in one's inside. Eating a raw onion a day would work, he argued, and it could be 'quite palatable' if a little salt was added. The onion could be sterilized by dropping it in boiling water for two minutes, first removing the outer scales. Apparently, the reason this worked was that 'onion contains allyl sulphide, a powerful germicide'. He also suggested chewing fresh parsley which could be sterilized in the same way as onions. Wild thyme and garlic were also used by his clients, and gargling with red sage was another option. He provided a recipe for a concoction of water, sage and vinegar. Although this was blatantly a form of free advertising in a letter to the editor, the editor also added at the end that another correspondent recommended a few grains of permanganate of potash in drinking or cleaning water.[135] More letters included concerns about cesspools, the suggestion that new tenants in houses which had been empty a while should run their taps well, and that information on the transmission of contact cases should be distributed, rather than the 'hush, hush' approach.[136]

Although reports in the tabloid and the left-wing press were sensationalist, the Chamber of Commerce and Ministry of Health were quite successful in their attempts to regulate representation of the epidemic within the press and used it to their advantage. The press could also be utilized to stimulate discussion of blame and responsibility leading to the public inquiry and calls for compensation.

Compensation in Croydon

Considering that English law is based on precedent, it appears from the typhoid test case that this was a landmark medico-legal case in England, with 260 compensation claims to a local authority for causing ill health.[137] Cases cited as precedents include a person injured by a falling tombstone, a snail in ginger beer, and lack of medicines on a ship, amongst other more directly related cases such as the use of lead pipes in a soft water area, and the use of the Food and Drugs (Adulteration Act) of 1928 in order to try to issue a penalty to a dairy in Bournemouth which sold milk containing the typhoid bacilli.[138] However, negligence was dismissed in the judgment in this latter case, whereas in his report on Croydon, Murphy provided ample justification for claims of negligence. He argued that the supply from the well could have been cut off whilst work took place, or another chlorinator could have been installed at the reservoir. Both of these actions took place following the outbreak.[139] Murphy claimed that if Holden had known that filtration and chlorination had ceased whilst men were working in the well, he would have provided advice which would have prevented the epidemic. However, not even Boast, the borough engineer, realized that the water would cease to be chlorinated during the work period. The men working on the well were new to the job, because previous employees had refused work after a man fell down the well and died in January 1937. Murphy suggested that if a medical examination of the new workers had been carried out, the carrier might have been investigated more closely as he looked unhealthy. He also commented that although bacteriological tests of faeces and urine would not usually be expected, Widal tests would not have been inappropriate considering the work on the well was not urgent.[140]

Following the report, the Water Committee offered to resign but the members of Croydon's local government decided that prevention, not punishment, was important in cases like this and that a committee would be established to consider the inquiry report.[141] The local residents were not so accepting. On 1 March, in response to the publication of the inquiry report, Rimington led a deputation to Croydon Corporation. He demanded, on behalf of the people of

Croydon, that a fully qualified water engineer should be appointed, the town's water should be bacteriologically examined daily, there should be cooperation between departments, local doctors should be appointed to water and public health committees, and there should be a medical committee. Rimington also proclaimed that those responsible for the water supply in October 1937 should resign, all who failed in their duty should be dealt with and compensation should be considered. These eight points were recorded across the national press.[142]

It was not novel for patients to assert rights in a court of law. In the eighteenth century, patients could successfully take their doctors to court for damages, to avoid paying unjustifiably high bills, or if treatments were unsuccessful or inappropriate.[143] More organized claims against medical negligence came with workers' rights, as shown in Chapter 3. There are previous examples of arguments for compensation for typhoid sufferers, including the settlement in Maidstone, as seen in Chapter 5.[144] If cases like this were settled out of court they would not enter the statute to be used as precedent in future legal cases. In Croydon, writs were issued to the local authority for personal suffering, loss of relatives and loss of trade.[145] Frank Furedi and Tracy Brown discuss compensation culture as a 'recent phenomenon' in Britain, especially how compensation claims currently drain the public purse. Yet the Croydon case shows there is a longer history of significant compensation claims.[146]

Furedi and Brown state that one of the reasons that Britain has been considered to be behind the US in its amount of litigation is that class action has only recently come into existence.[147] The test case, in which two of the plaintiffs claimed compensation following the Croydon epidemic, demonstrates that although mass claims have not been put forward in a consolidated form as in the US, the use of a test case in Britain can have the same purpose, so a form of class action was in effect in the 1930s in Britain.[148] In the judgment it was stated

> The present case is a Test Case, inasmuch as it has been agreed between the Corporation and certain other persons who also contracted typhoid fever, and who have taken proceedings against the Corporation, that the findings of fact and law so far as they are applicable shall be treated as decisive of these additional claims.[149]

A recent leaflet published by the Accident Line of the Law Society tells the reader, 'sometimes you don't even realize that someone or something else is to blame'.[150] This type of attitude is shown when analysing the Croydon epidemic, so finding someone to blame is not new – the idea that even if the sufferer of an accident believes that there was no one to blame for an incident, they may find otherwise upon talking to a solicitor.[151] Writs began to be issued to the Council in March 1938. A bill had to be passed in Parliament in order to allow people to claim who did not issue writs within six months of their illness, suggesting that people could jump on the bandwagon in order to claim money from the Corporation.[152]

In addition, the *Daily Mail* reported that a member of the Croydon Council of Social Service, who was also a partner in a solicitors' firm, had 'touted' for trade amongst typhoid sufferers whilst in his role as a social worker.[153] In March 1938 the Corporation estimated the cost of compensation would be £4,511 (1937 sterling).[154] However, in December Councillor Maycock estimated that the people aiming to claim compensation could cost the council £50,000 to £100,000 as hospital bills alone had amounted to at least £20,000.[155] By November 1939 the payments had indeed amounted to £92,168.11s. 2d. (£3.4 million in 2005 sterling)[156] resulting in the Corporation having to borrow this money from the Ministry of Health, ultimately being paid for by the ratepayer.[157]

Excessive claims are not a new phenomenon either, and were made in the test case. Alfred Read attempted to claim for a hotel bill of £63.11s.0d. (£2,313 in 2005 sterling) from Croydon Corporation. He stayed with his daughter in a hotel whilst another daughter was sick, even though he lived in a house of thirteen to fourteen rooms, including bathrooms. Although nurses came to live at his house during his daughter Patricia's illness, it was still judged that there was plenty of room.[158] Although the Outbreak Committee had organized the citizens of Croydon, it claimed that the aim had not originally been compensation, but that they had heard that people were contemplating claiming if the inquiry showed justification.[159] These residents of Croydon felt that the water was polluted and they wanted to know who was to blame – both for allowing the water to get polluted and for not disseminating information to the public sooner. Vanessa Taylor and Frank Trentmann state that consumer rights in general began with complaints about the water supply, and their study on Victorian London focuses on the value, supply and need for water for personal hygiene in an increasingly civilized society, rather than complaints about water carrying disease.[160] Murphy questioned Holden, in court, demonstrating the lack of official means to complain about the bad quality of water in 1937:

> Water is a thing that normally one has to take and one has to pay for, and if you do not pay for it, as it is on your rates, you are liable to penal consequences. I am not suggesting it in this case at all, but assuming somebody is selling you water that is full of bacillus typhosus, that person is under no penal consequences at all?

The doctor replied, 'That is how I see it.'[161] The consumer was finally given rights following the epidemic. The test case brought against the Croydon Corporation in the autumn of 1938 resulted in compensation for medical expenses and 'pain and suffering and general inconvenience caused by [the plaintiff's] illness'.[162]

The Croydon case, with its 260 claims, appears to have been extraordinary, and yet it is not recorded in the literature on disease and compensation.[163] Studies of legal cases in relation to negligence in looking after health have focused on employees, traumatic accident victims and children.[164] Literature on illness and law concentrates on occupational health, and Jane Stapleton argues that the chances of gaining compensation for infectious disease are very unlikely, except for anthrax. Conditions of tort are proof of the cause of illness and the establishment of the time at which the disease was caused. In addition, the individual, hospital or authority at fault must be pinpointed and evidence must be provided that the risk of disease should have been known and that the defendant's response was inadequate.[165] The public inquiry provided the plaintiffs with the ammunition they required to claim compensation from the Corporation, and the requirements of tort law demonstrate that the pinpointing of a carrier was important for claiming compensation. The Croydon test case should be iconic in legal histories in demonstrating how a successful compensation claim can be made for an epidemic.

Blame did not stop with monetary compensation claims in Croydon, with Holden becoming the scapegoat for the community's anger. According to the *Sunday Dispatch*, Holden became the 'most harassed man in Britain' and was provided with a constant police guard at his house after he received some threatening letters and callers. Holden told the newspaper, 'After a certain time, one can get used to almost anything ... Even when everyone seems to be turning against you.'[166] There was a pre-existing dispute between the local GPs of Croydon and Holden regarding maternity services.[167] Dr Forbes, appearing at the inquiry on behalf of the Croydon Division of the British Medical Association and the Croydon Local Medical and Panel Committee, questioned Holden about cooperation.[168] He took the opportunity to criticize the composition of the Public Health Committee, which included no 'medical men', as members were elected. He also condemned Holden for attending the residents' meeting of 'laymen' in order to gain their opinion, but not arranging a meeting of local doctors to gain theirs.[169] Forbes even attacked Holden with the question 'You are out of touch with private practice, are you not?'[170] Following Forbes's heated questioning of Holden, Murphy asked Forbes to 'bear in mind ... the distinction between collaboration generally in the absence of an outbreak, collaboration between the MOH and the local medical profession, and collaboration after an outbreak has taken place'.[171] In addition to criticism by residents and the press, Holden had to endure questioning by the local medical profession, presumably inflamed by their ongoing dispute. Yet, if the press reported correctly, Forbes was hypocritical as he was quoted in the *Daily Telegraph* on 25 November as saying, 'As a purely personal opinion, I should like to say that I think a mountain has been made out of a molehill in this outbreak. It has been called an epidemic, but it is little more than an outbreak.' This comment apparently followed a meeting with 300 Croydon doctors on 24 November.[172]

As shown in Chapter 5, prior to 1937 charity was the usual way in which help was provided for hardship caused by an epidemic. In Croydon, Rimington requested donations in a letter to *The Times*, in which he appealed to the nation. Yet, rather than appealing for funds for those who were suffering from the epidemic, he appealed for funds from 'public-spirited individuals' to enable the South Croydon Outbreak Committee to have legal representation at the public inquiry regarding the epidemic. However, he did mention that any surplus would be donated to the Croydon General Hospital.[173] In February 1938, he wrote again to *The Times*. Sums of money and offers of employment for victims of the epidemic had been sent to him and he had handed them to the Bishop of Croydon. He asked for any further offers of help to be sent directly to the bishop.[174] The Typhoid Relief Fund raised £296.13s.6d. by 2 March 1938, far from £8,700 raised through the relief fund for the Malton epidemic.[175]

Perhaps the increasing role of the National Health Insurance scheme and state-funded hospitals hindered actions of charity by the late 1930s, with ideas that the state was responsible for the cost of regaining health. Indeed, Martin Gorsky, John Mohan and Martin Powell have shown that charity declined in the financing of hospitals in the 1930s, and the growth in municipal funding and taxation for this purpose paved the way for state funding of health care. Before a decline, the peak in the financing of hospitals through charitable funding was reached in 1932, the same year in which money was raised for victims in Malton despite problems of the depression and unemployment.[176] Therefore the response in Croydon may have been as a result of an increasing reliance on the state to look after victims of disease. Indeed, this trajectory is demonstrated by a decision during a typhoid epidemic in 1964. The idea of a relief fund for sufferers of typhoid in Aberdeen was rejected at a special council meeting.[177]

The social class of the residents of South Croydon may have affected their response to the epidemic. The focal area for the residents who campaigned regarding the typhoid epidemic was centred around St Augustine's Avenue in South Croydon. This was in a middle-class area, demonstrated by comparisons with rate valuations in various areas of Croydon. Sales records for a property with similar valuations to Rimington's and Green's on this road show that they probably lived in three-bedroom houses with an additional box room, and houses with higher valuations on that road benefited from an extra study and scullery.[178] Richard Rimington went to the private Whitgift School. Judging by the information available, these were very different characters from the woolsorter-activists in Bradford, for example. Alfred Read, the secretary of a big industrial organization, was one of the plaintiffs in the test case. He employed a maid and lived in a house with a rate valuation of £136 compared to Rimington's at £48. Rimington's and Green's houses had a median average valuation for the area, whereas some roads had houses with typical valuations of £68 to £101.[179] Therefore the leaders of the South Croydon Outbreak Committee were not the wealthiest citizens of Croydon, but middle-class citizens directly affected by disease.

The carrier working on the well in Croydon could have been blamed by the community, since he was pinpointed as the origin of the epidemic in the inquiry. However, the representative of the Ministry of Health was quick to assert that the carrier was not to blame – after all, 'he was a man who had served in the war, had contracted typhoid in the war, and, through no fault of his own and in ignorance of the fact himself, remained a carrier of the bacillus'.[180] Although carriers were sought and usually discussed anonymously in the press as causes of epidemics from 1905, they were not found to be culpable. Even Antonio the ice-cream man from Eccles, named in 1910, was not blamed, despite his condition contributing to the death of eight people.

Lay people in Croydon were active, had knowledge of bacterial disease and they were keen to apportion blame. Comparing this case with earlier epidemics helps to explain how this knowledge and response developed. As demonstrated in Chapter 5, blame and responsibility were rife as early as the 1882 Bangor epidemic, but in a very different form to the Croydon epidemic. In this instance, rather than the local community, it was the MOH and the local medical profession who argued.

More recently, in Malton in 1932, ideas of compensation were increasing. People wrote to the charitable fund organizers to ask for compensation. The way in which typhoid was transmitted in Malton was similar to transmission in Croydon, and yet did not provoke the same levels of anger and desire for retribution. A scapegoat was not sought in the same way as in Croydon, but there had been some calls for an inquiry, and complaints about complacency and the length of time in sorting out the water supply which led to extra damage to trade.

There are several reasons as to why the anger generated by the Malton epidemic did not reach the same levels as that caused by the Croydon epidemic. Even though a surgeon died in Croydon, he does not appear to have been directly involved in selflessly treating outbreak victims like George Parkin, the general practitioner in Malton.[181] The economic depression and the boycott of the town may have influenced people to work together. The extent of charity suggests this too. Also, in comparison to Croydon, there may have been a different community oriented mentality in the small Yorkshire town, compared to a London commuter town like Croydon.

The success of the Croydon compensation claims impacted on attitudes towards blame and responsibility in the subsequent typhoid epidemic in Aberdeen. In 1964, the Treasury prepared documentation on the legal position, and stated that it was unlikely to succeed. It would be too difficult to deal with compensation from Argentina, the source of the contaminated tinned meat, and the manufacturers and businesses losing trade as a result of the epidemic were too far detached from the origin.[182] The Treasury claimed that in the Croydon case, the local government had been negligent and that this was why compensation was awarded, whereas in Aberdeen there was no negligence by government. Lady Tweedsmuir, Conservative MP for South Aberdeen, said the 'whole of Aberdeen' was seeking compensation, comparing the epidemic with floods and disasters for which compensation was given. A total of £7 million in trade had been lost. However, in Aberdeen, only a small amount of money was paid by the hospital board and this was for damage to belongings by steam sterilization.[183]

Fear and Flight

Although many people in Croydon were aware of disease transmission and bacteriological testing, older ideas still permeated reactions to bacterial disease. Fear and flight continued to be a problem in schools. At Richard Rimington's independent Whitgift School, only three boys were away from school on 20 November 1937, although it was announced that 'any parent has the right to keep his son away'.[184] By 27 November, 128 boys were away out of the school's 700 pupils.[185] The number absent declined gradually from this date, but the sudden increase showed the panic which was gripping the town. A school notice tried to combat this absenteeism, stating 'Everyone who has been into Croydon is just "in the boat."'[186] The proportion of students staying away from the school was very similar to that at Marlborough College in 1890, as seen in Chapter 5.[187]

In the test case, Alfred Read tried to claim that flight was natural in a time of epidemic. In explaining his claim for hotel expenses, his lawyer, H. J. Wallington, stated 'it would be unreasonable for anyone who could get away to stay in an area where it was perfectly obvious that typhoid was raging'. When Justice Stable asked him, 'Do you say that the entire population of Croydon would be justified in departing from Croydon and going to live somewhere else until the danger was removed', Wallington answered, 'I should go as far as that, my Lord. I know quite well that if I lived in Croydon and could get away I should fly at the earliest possible moment. One could never tell who would be next.'[188]

Read was questioned regarding his reasons for moving to the hotel with his elder daughter. He argued that his maid was also ill, and that it was impossible to find more help as no one would come to his 'typhoid infected house'. He talked to his wife, and decided that he and his daughter would move out of the house, later clarifying that this was because the maid and his wife would not be able to attend to him and his elder daughter as well as to the nurses and his younger daughter. Read also argued that as the secretary of a 'big industrial organisation' he did not want people to worry that contact with him put them at risk of infection. On further questioning, he revealed that neither he nor his daughter was allowed to the office or school even after they had left the house, but Read felt that he 'was much handier in London'. Monckton asked whether Read's colleagues were aware that typhoid was not communicable by breath, but

Read argued that if a member of staff had contracted typhoid it might have been said that he should not have been in the office. He continued, 'even amongst my friends who knew that my daughter had typhoid I was avoided by them whilst I was in the Hotel'. Read was also challenged as elementary schools were not allowed to keep children away because of typhoid, but he argued that his daughter was 19 and was told not to return to the Central School of Speech Training and Dramatic Art, based at the Royal Albert Hall in South Kensington, for the incubation period of three weeks.[189] Monckton further established the Corporation's defence regarding the hotel bill by asking Eyre to confirm that typhoid cases were still treated on general wards in hospitals whereas measles cases were treated in a fever hospital. Eyre argued that ideas about fever transmission were changing, with measles and scarlet fever being treated in the same wards.[190]

George Buchan, MOH for Willesden, was also presented as a witness and was asked whether it was necessary for Read and his daughter to leave his fourteen-roomed house. Buchan answered that it was unnecessary as the patient could have been isolated. He argued that there had been five cases of typhoid nursed at home in the last six or seven years, all in smaller houses, and some more densely occupied, and there was no spread of disease to other family members in any of these cases. He also proposed that the younger daughter could have been treated in hospital.[191] Considering the maid's illness, Justice Stable commented that the elder daughter could have made her own bed, and helped with the domestic work of the house.[192] With the judgment being that Read would not get his hotel expenses, the court was convinced by medical recommendations that flight was not necessary if proper precautions were taken to prevent the spread of bacteria.[193]

The increasing avoidance of public gatherings over the period of this study continued with the Croydon epidemic, despite Horder's protestation. The lack of people on the streets on a Saturday was noted in the *Sunday Referee* on 21 November 1937:

> Many people, especially those with children, hardly dare to leave their homes, fearing infection if they venture among crowds. North End and George-street and other principal thoroughfares were almost deserted yesterday. One shopkeeper said to me: 'It has been my worst Saturday since the war. People seem to dread rubbing shoulders with other shoppers.'[194]

In spite of this general avoidance of public gatherings, including absenteeism from school, competitive inter-private school sport seems to have been perceived as important enough to continue. The Whitgift School captain gave instructions on 18 November: 'All schools playing any Whitgift Team must be notified.' Visiting schools were informed of the epidemic before their rugby teams went to play there.[195] However, the matches still continued, with teams from other schools coming to Whitgift, indicating that other schools believed that these matches were important enough to risk their health, or that they did not believe that typhoid could be caught through playing this contact sport.[196] The historian of the Whitgift School comments on the importance of sport to elite schools during the 1920s, namely *cultus athleticus*:

> [the] back pages of the *Observer* and *The Sunday Times* ... filled with short reports of inter-school matches. The secretary of rugby was required to send to an agency immediately after a home game a shilling telegram giving the score and a sentence or two about the play.[197]

The importance of sport in elite English education was a tradition. As noted by Andrew Warwick, it became 'characteristic of British educational establishments' in the mid-Victorian era, with ideas of 'a more liberal notion of Christian manliness' and improvement of discipline, health and appetite, as well as attributes such as competitiveness and teamwork.[198]

Even though bacteriology had developed into a strong research and service discipline by the 1930s, many responses had remained the same since 1880. Flight and attitudes towards public gatherings continued. Perhaps more bacteriological knowledge inspired a calmer approach for some people, but caused others to panic more. Heather Prescott states that increased knowledge brought increased fears, so perhaps more knowledge of the dangers of bacteria encouraged people to fly from schools, and not to go into towns, even though notices told them that with hygienic precautions they would be fine to do so.[199]

Questioning Expertise

The trained experts on infectious disease and the water supply were the MOH and the borough engineer, but their authority was challenged and usurped by the citizens of Croydon. Freda Churchill explicitly questioned their judgement in her letter to the *Croydon Advertiser* on 13 November, mentioned earlier.[200] The Political and Economic Planning report written in the same year as the Croydon outbreak stated, 'The most important administrative medical appointment under a local authority is that of medical officer of health ... Obviously the work of a medical officer of health is very responsible and requires considerable talents and a detailed training.'[201] However, it was Rimington who was praised within Murphy's report:

> the conduct of Mr Rimington ... seems to me to have been that of a very public-spirited citizen and that his emphasis on the question of water was not only justified in the event but may have been of considerable value.[202]

Earlier disease outbreaks, discussed in Chapter 5, have provided examples of when doctors' authority was undermined. For example, the city administration in Bangor chose to ignore the MOH and most of the doctors. Despite the increasing complexity of bacteriological tests in the 1930s, public understanding of disease transmission led to the citizens of Croydon's incredulity that so-called experts such as the MOH and the borough engineer could have allowed the epidemic to occur, leading to the subsequent compensation claims. Not only had those with positions of authority in the local corporation failed to spot the cause of the epidemic, but they had also failed to make the water safe. Yet it was only following the Croydon epidemic that chlorination was continuously used for most public water supplies.[203] As Mendelsohn points out, knowledge of carriers led to a focus on individual people rather than 'swamps, streets, and sewers' in the early twentieth century.[204] However, there was a turn to a more holistic and ecological perspective in epidemiological and bacteriological research in the interwar years.[205] In the 1930s in Malton and South Croydon, the focus of responsibility concerned sanitation and the environment, not the individual. Local communities blamed local and national governments for allowing typhoid bacilli emitted by a patient and a carrier to enter the water supplies.

CONCLUSION

I have consulted with Dr Walker, + Dr MacNaught of York, + neither of them seem to be able to suggest any further method of treatment for my husband. It seems incredible that with medical science at such a pitch as it is today, that a case should go on indefinitely like this. Surely there must be some cure?[1]

Mrs H. to Malton Typhoid Relief Committee, 2 May 1933

In 1933, Mrs H. expressed her dismay that her husband was not recovering from typhoid. Sixteen years before the antibiotic chloramphenicol was found to successfully treat typhoid, Mrs H. could not understand why modern medicine was not capable of healing her husband. Bacteriology brought significant changes to the experience of diphtheria patients in hospital, and increasingly for people suffering from anthrax, but there was also considerable continuity in medical knowledge and practice. A short review of continuity and change in a range of contexts in various locations will consolidate the research presented in *Bacteria in Britain*, comparing the value of bacteriological technologies and knowledge in the hospital, the workplace and the community in the late nineteenth and early twentieth centuries. But first, the conclusion addresses this volume's significant revision of the historiography of the laboratory and clinic.

Chapters 1 and 2 contribute to the historiography of the social history of medicine but also to wider themes in the writing of the history of Britain. The historiographical trend towards declinism and nostalgia in Britain appears to have influenced the secondary literature about the relations between the clinic and the laboratory. Peter Mandler criticizes the concentration on rural nostalgia within the recent Englishness genre, following historians' belief in the decline of Britain formulated in the 1970s, which has increased since Martin Wiener's *English Culture and the Decline of the Industrial Spirit 1850–1950* in 1981.[2] This present-day nostalgia appears to have influenced the idea that gentlemen physicians were opposed to new technology, focusing on the individualism of the patient.[3] Ethicists may also have influenced nostalgic discussions in light of the perceived decline of bedside medicine, resulting from current concerns about 'the infringement of technology on the relationship between doctor and patient'.[4] Additionally, as discussed in Chapter 2, there are still concerns about communication between the laboratory and the clinic today.

The historiography of the 'gentleman' has especially influenced current representations of physicians at Bart's. The staff at Bart's in the late nineteenth and early twentieth centuries included doctors who were physicians to monarchs and prime ministers as well as a president of the RCP. Bart's had a much larger percentage of public-school-educated physicians than the Fellows of the RCP as a whole.[5] Marcus Collins has analysed the attitude that after the 1950s, gentlemanliness, epitomized by the public school, Oxbridge and sportsmanship, was responsible for the decline of Britain.[6] Collins proposes that the English gentleman was used as a 'scapegoat' at a time of relatively slow economic growth and decolonization.[7] It appears that this idea of a gentleman's character has also been used as an explanation for their attitudes to the laboratory, instruments and specialization.

Additionally, sociological concerns about competition between professional groups may have directed the emphasis on historical tensions in the laboratory and clinic relationship. Steve Sturdy has proposed that the history of the professionalization of science and medicine has emphasized the 'pursuit of professional autonomy, authority and ultimately power'. As with the growth in declinist literature, this change also began in the 1970s. Sturdy argues that the relationship between science and medicine was a subject of debate at the time, with one side arguing medicine was not scientific enough and needed to be scrutinized, and the other seeking to preserve 'patients' rights and dignity' through preventing medicine becoming overly scientific. Therefore these modern concerns were projected back to the late nineteenth century by medical historians.[8]

Following both of these historiographical trends, histories of Bart's changed emphasis in the 1970s, as discussed briefly in Chapter 1. In 1918, Sir Norman Moore, a gentleman physician, wrote an immense history of Bart's, which examined the opening of many hospital departments in detail, including pathology.[9] He chronicled the increasing staff of the department, its growing buildings and the contribution to the 'daily work of the hospital', writing that '[t]his development shows the living vigour which St Bartholomew's possesses in the eighth century of its existence'.[10] Since the 1974 volume, which portrays a conservative late nineteenth-century hospital, was written by doctors, this influence of declinism, nostalgia and interprofessional tensions can be seen within the medical community as well as through historians since the 1970s.[11]

Suggestions have been put forward within *Bacteria in Britain* to explain why the contemporary and historiographical representations of the relationship between the laboratory and clinic are complicated. First, language was conservative at Bart's, indicating that the physicians may have been torn between the old world of bedside medicine and the new world of the laboratory, holding on to the old world by continuing to use its terminology. Second, Samuel Gee was a complex character, and did not wholeheartedly accept anything, using new knowledge or practices on his own terms. He also seems to have accidentally presented an image to students that he was somewhat opposed to the laboratory, and this image of Gee has been perpetuated ever since. Thomas Horder was at first wholeheartedly enthusiastic about the laboratory and the specialists who worked there, but began to question this in the 1930s when his status and income were challenged. The change in the historiography of Bart's may therefore align with the declinist and nostalgic writing on the history of England in general, the sociology of professional tensions and may also be due to ethical and practical concerns regarding the value placed on laboratory results in comparison to bedside medicine.

Although Chapters 1 and 2 have demonstrated a rapid adoption of bacteriological diagnosis for diphtheria and typhoid, the use of the laboratory was not as regular for pulmonary tuberculosis, for which pre-existing clinical examinations were already relatively useful in detecting the disease. For all three diseases, and especially pulmonary tuberculosis, the test results from the laboratory were not infallible. Therefore, although the use of the laboratory became routine, this change did not constitute a revolution; existing methods continued to be utilized, and bacteriological tests were usually performed in order to confirm the results gleaned from clinical methods of diagnosis.

For the past twenty years, Michael Worboys and Nancy Tomes have dispelled myths that a bacteriological revolution in medicine resulted in a fissure in concepts of disease, focusing respectively on how bacteriological knowledge was researched, and on how it was disseminated and applied in everyday life.[12] Just as there was continuity in diagnosis and terminology in the hospital alongside changing concepts of disease, Chapters 3, 5 and 6 have demonstrated continuity in how communities reacted to disease. People sensed the risk of disease from the presence of bad smells; those who could afford to, continued to flee from epidemics; and most lay methods of protection from disease within the home were largely derivative of traditional hygiene methods or the sanitarian movement. Chapters 5 and 6 have in particular revealed the importance of the older science of epidemiology within both medical and lay communities, for which bacteriology was a useful confirmation. On the other hand, concerns about chlorination, carriers, inoculation and flies stemmed directly from bacteriology. One of the most contradictory discoveries within Chapters 5 and 6 are the different reactions to chlorination and inoculation, direct products of experiments with bacteria. With only five years separating the Malton and Croydon epidemics, the citizens of Croydon were horrified that the water had not been purified, yet the Yorkshire community believed that they did not need chemicals in the water but sought a pure source from the local rural environment. The opposite reaction occurred with regard to inoculation. Whereas many of the residents of Malton appear to have actively sought inoculation, it was avoided to an extent in Croydon, following medical fashion at the time.

While changing concepts of typhoid in the community certainly did not constitute a bacteriological revolution, the speed with which physicians, lawyers and employees used bacteriological test results and concepts of disease illustrates that in some circles the science made an immediate and distinctive impression. The techniques and knowledge of the new science of bacteriology were quickly absorbed in the hospital and in industrial disputes in the 1880s and 1890s.

Within the hospital, the discovery of a cure for diphtheria appears to have been crucial for the introduction of clinical laboratories and pathologists at Bart's; the fast reception of the Widal test for typhoid quickly followed, even though there was no cure for this disease. The importance of diphtheria antitoxin for the reception of practical bacteriology at Bart's is highlighted by the comparison with the slower adoption of bacteriological diagnosis at Addenbrooke's Hospital, where the removal of patients with diphtheria to a fever hospital may have affected the reception of laboratory tests. Diphtheria was also important because of public enthusiasm about the discovery of antitoxin, an argument used to persuade the treasurer at Bart's to provide funding for the laboratory. Indeed, public support had a huge impact in France, as Paul Weindling has discussed, leading to 'lavish' donations towards serum production.[13] Although diphtheria was of great significance at Bart's, the financial response from the local community was not as enthusiastic due to the hospital's reputation of being enormous wealthy.

Despite anthrax and typhoid having very different modes of transmission, and requiring different prevention and management strategies, the concept of a microbe causing disease transcends this, and this idea was utilized at different rates by lay people in order to argue for safer environments. Bacteriological theories were adopted quickly in the Bradford workplace and led to a change in treatment methods at Guy's Hospital for anthrax in Bermondsey. In both Bradford and London, medical and popular knowledge of the aetiology of anthrax was radically changed by bacteriology – whether in reframing woolsorters' disease as internal anthrax, or understanding the reason for malignant pustules which were previously thought to be caused by chemicals for curing hides. This did not necessarily lead to a change in practice in the workshops as rules were based on sanitarian ideas of cleanliness.[14] However, these rules were reinforced by the confirmation of the bacterial aetiology of the diseases. The definition of these diseases as bacterial also led to increased interest from the state, for example, the LGB sent John Spear to investigate. The high-profile response to anthrax in Bradford may have resulted from socialism and the well-publicized discoveries of Koch, and in particular Pasteur, who was also directly linked with treating Bradford residents for hydrophobia. On the other hand, it was sometimes difficult to prove cases of disease. A similarity between anthrax and typhoid is that it could be difficult to find the pathogenic bacteria. The typhoid bacillus could be difficult to detect in water during an epidemic and, as was illustrated by the autopsy of Indiarani the elephant, the bacteria associated with putrefaction could overcome the anthrax bacillus, removing the evidence of the cause of disease.

When available, proof of the cause of the disease through laboratory experiments certainly lent more credence to complaints at inquests and for compensation claims. Legally, in both the workplace and the community, bacteriology was essential for providing evidence of how a disease was transmitted and why people died. Mass compensation for disease demonstrates the importance of isolating bacilli for legal cases, as well as the evolution of ideas of responsibility for bacterial disease from the early Workmen's Compensation Acts to the community compensation claim in Croydon. Indeed, compensation had been sought for typhoid in Maidstone in 1897, the same year as the first Workmen's Compensation Act. Evidence from letters written during the epidemic in Malton suggests that compensation culture was developing further during the early 1930s. However, although workmen's compensation was given using a no-fault principle, negligence had to be proven in order for a compensation claim to be successful for a typhoid epidemic.

Bacteria in Britain has also demonstrated that reactions to the same diseases were often very different depending on the location, furthering the current trend in considering the importance of place for histories of medicine.[15] The response to bacteriological concepts of disease was significantly influenced by local places and policies. In turn, as Charles Rosenberg has illustrated with his study of cholera in New York, studying the history of disease can be very useful for highlighting broader social, economic and political histories of a particular place.[16] In the case of east London, the plight of the leather workers did not receive the same national recognition as that of the woolsorters in Bradford. The industrial politics which led to woolsorters being respected and highly paid, while dockworkers suffered underemployment and job insecurity, were also highlighted by the government's concentration on finding a solution to eliminate anthrax from wool rather than leather. During the first thirty-five years in which anthrax was defined as bacterial, there was no more evidence that a successful solution for disinfection would be discovered in wool than in hides and possible chemical treatments were proposed for each material. The reasons why the risk from wool were taken more seriously than the danger of leather lie outside contemporary medical knowledge and probably result from the prominence of the wool trade in Bradford compared to the relative anonymity of workers involved with leather in London, pointing to the importance of advocacy.

In London, Guy's Hospital doctors and bacteriologists were very involved with trying to treat anthrax amongst the local community. However, successful advocacy was apparently dependent on a combination of forces: employees, employers, unions, the medical profession and the state. In addition to workers being much more concerned about issues such as underemployment, perhaps the relative lack of recognition nationally was because there was local support; anthrax gained significant recognition locally in the nearby hospital and within the leadership of the leather trade, with *Leather World* showing concern for the plight of employees and the Worshipful Company of Leathersellers's employment of a bacteriologist to research the problem. Examining the contexts of both medical and lay knowledge within this book has revealed that in the early twentieth century, medical interest alone was not enough; heated exchanges and debate between and within groups of employers, employees and doctors were perhaps crucial in gaining publicity.

The working-class experience of politics about disease in the capital was entirely different from that of the middle class. The clamour for work in poverty stricken east London led to a relative lack of publicity about anthrax which perhaps resulted from the anonymity of the working-class dockers and tanners in London. In contrast, the Croydon epidemic was infamous in the national press probably because of its proximity to London. Since Croydon is only ten miles south of London, Charles Rimington was able to attract considerable national interest in the disease. His ability to easily visit the Ministry of Health for a deputation on the same day as continually carrying out his own investigations into the disease in Croydon led to experts from London being called to an inquiry in Croydon whilst the epidemic was still in full swing – a unique response. Additionally, the publicity of the risk to people living in the capital has been illustrated by cases of typhoid in Kensington, an employee of Lloyds in the City falling ill in her office, and fears that watercress sold in the metropolis was responsible for the disease. Yet in a rural town in North Yorkshire a much more stoical response was seen, with the deputation to the Ministry of Health being organized for a completely different purpose, involving local council workers who were concerned about the costs of the epidemic and the provision of a new water supply. The importance of class has also been illustrated by the response to the contamination of shaving brushes. This international incident could affect anyone, rich or poor, who intentionally or accidentally bought a shaving brush with imitation badger hair. This led to urgent notices in the press and, perhaps more than any other case study examined within this book, alerted a significant proportion of the national population to the dangers of bacteria and methods of disinfection.

Returning to Mrs H.'s lament, what was the point of bacteriology? In the case of typhoid, from which her husband suffered, bacteriological diagnosis was routine but how was this of value when treatments were not very successful? Howell has pointed out that prognosis was still important: 'being able to predict a fatal or a benign course was of more than casual interest to the patient and his or her family'.[17] Bacteriology was also important for diagnosis when fighting the further spread of the disease, as has been shown in Malton, Bournemouth and Croydon, for example. The allied science of immunology had also developed an inoculation for the disease which had a variable reception in times of epidemic. Within the hospital, the development of diphtheria antitoxin has been shown to be crucial for the funding of the bacteriological laboratory and was obviously of great importance for victims of the disease. In Bradford and London, bacteriology was valuable for defining what woolsorters' disease and malignant pustules were, and to enable research into prevention of anthrax. Lawyers certainly saw the significance of the results of bacteriological tests as evidence for compensation claims for both workplace and community diseases. Yet, despite evidence of a dramatic change in practice and procedure accompanying the development of the bacteriological laboratory, the analysis of a variety of themes, contexts and diseases has revealed that the broad reception of the new science was not so simple. Even though it appears that there was not a revolution in medical and lay practice or beliefs following the discoveries of Pasteur and Koch, bacteriology was certainly of sustained and broad use to elite physicians, lawyers, employees and concerned local communities in Britain.

NOTES

Introduction

1. A substantial amount of bacteriological research was not concerned with disease in humans and animals, although bacteriology has generally been associated with infectious disease. See J. A. Mendelsohn, 'Cultures of Bacteriology: Formation and Transformation of a Science in France and Germany, 1870–1914' (PhD dissertation, Princeton University, 1996), p. 13. Refer to K. Vernon, 'Pus, Sewage, Beer and Milk: Microbiology in Britain, 1870–1940', *History of Science*, 28 (1990), pp. 289–325, on pp. 305–15, for many examples of other research and analysis such as agriculture and brewing.

2. See R. Bud, *Penicillin: Triumph and Tragedy* (Oxford: Oxford University Press, 2007).

3. See M. Worboys, *Spreading Germs: Disease Theories and Medical Practice in Britain, 1865–1900* (Cambridge: Cambridge University Press, 2000) for continuity of theories of disease amongst the medical community until 1900. See N. J. Tomes, *The Gospel of Germs: Men, Women, and the Microbe in American Life* (Cambridge, MA, and London: Harvard University Press, 2002) for continuity of public perceptions of disease transmission until the 1920s in the United States of America.

4. For example, Worboys, *Spreading Germs*; Vernon, 'Pus, Sewage, Beer and Milk', pp. 89–325; A. Hardy, 'Food, Hygiene, and the Laboratory: A Short History of Food Poisoning in Britain, *circa* 1850–1950', *Social History of Medicine*, 12 (1999), pp. 293–311; C. Hamlin, *A Science of Impurity: Water Analysis in Nineteenth-Century Britain* (Berkeley and Los Angeles, CA: University of California Press, 1990). Older general studies of bacteriology include W. Bulloch, *The History of Bacteriology* (London: Oxford University Press, 1938); W. D. Foster, *A History of Medical Bacteriology and Immunology* (London, William Heinemann Medical Books Ltd, 1970).

5. B. S. Gregory, '*Is* Small Beautiful? Microhistory and the History of Everyday Life', *History and Theory: Studies in the Philosophy of History*, 38 (1999), pp. 100–10, on p. 105; H. Medick, '"Missionaries in the Rowboat"? Ethnological Ways of Knowing as a Challenge to Social History', in A. Lüdtke (ed.), *The History of Everyday Life: Reconstructing Historical Experiences and Ways of Life*, trans. W. Templar (Princeton, NJ, Princeton University Press, 1995), pp. 41–71, on pp. 47–8; S. Lindqvist, 'Change in the Technological Landscape: The Temporal Dimension in the Growth and Decline of Large Technological Systems', in O. Granstrand (ed.), *Economics of Technology* (Amsterdam: Elsevier Science B.V., 1994), pp. 271–88; D. Edgerton, 'From Innovation to Use: Ten Eclectic Theses on the Historiography of Technology', *History and Technology*, 16 (1999), pp. 111–36.

6. C. Gradmann, *Laboratory Disease: Robert Koch's Medical Bacteriology* (Baltimore, MD: The Johns Hopkins University Press, 2009), p. 7. This is also an approach taken by

Tomes, *Gospel of Germs*, which uses different types of case studies including bathroom ceramics and communion cups in order to understand the influence of bacteriology on American society.

7. See A. Irwin, *Citizen Science: A Study of People, Expertise and Sustainable Development* (London: Routledge, 1995), for the use of this term to refer to different groups of the general public.

8. B. Wynne, 'May the Sheep Safely Graze? A Reflexive View of the Expert–Lay Knowledge Divide', in S. Lash, B. Szerszynski and B. Wynne (eds), *Risk, Environment and Modernity: Towards a New Ecology* (London, Sage, 1996), pp. 44–83.

9. Irwin, *Citizen Science*, pp. 33, 51, 114.

10. Wynne, 'May the Sheep Safely Graze?'.

11. J. B. Fressoz, 'Beck Back in the 19th Century: Towards a Genealogy of Risk Society', *History and Technology: An International Journal*, 23 (2007), pp. 333–50.

12. K. Waddington, *The Bovine Scourge: Meat, Tuberculosis and Public Health, 1850–1914* (Woodbridge: Boydell and Brewer, 2006), p. 188; A. Woods, 'Foot and Mouth Disease in 20th-Century Britain: Science, Politics and the Veterinary Profession' (PhD dissertation, University of Manchester, 2002), pp. 123–4.

13. Examples of studies of bacteriology and the public in the USA include Tomes, *Gospel of Germs*; S. Hoy, *Chasing Dirt: The American Pursuit of Cleanliness* (New York and Oxford: Oxford University Press, 1995); D. Valenze, *Milk: A Local and Global History* (New Haven, CT: Yale University Press, 2011), pp. 210–32, briefly examines the role of the American consumer. V. Taylor and F. Trentmann, 'Liquid Politics: Water and the Politics of Everyday Life in the Modern City', *Past and Present*, 211 (2011), pp. 199–241, analyses the growth of British consumerism in relation to water, with groups such as the Stockwell and South Lambeth Water Consumers' Defence Association, but focuses on rates and cost, sanitary ware, water shortages and wasted water.

14. For quotes and discussion of AIDS activism, but also activism regarding breast cancer, chronic fatigue and more diseases, see S. Epstein, *Impure Science: AIDS, Activism, and the Politics of Knowledge* (Berkeley, CA: University of California Press, 1996), pp. 207, 219, 234, 346–53.

15. However, these postulates were first defined by Koch's research associate, Friedrich Loeffler, in 1883. See Worboys, *Spreading Germs*, p. 177.

16. Gradmann, *Laboratory Disease*, p. 109.

17. See W. W. Ford, *Bacteriology* (New York and London: Paul B. Hoeber Inc., 1939), p. 102, for many more examples.

18. Gradmann, *Laboratory Disease*, p. 5.

19. A. Hardy, '"Straight Back to Barbarism": Antityphoid Inoculation and the Great War, 1914', *Bulletin of the History of Medicine*, 74 (2000), pp. 265–90, on p. 273; Foster, *History of Medical Bacteriology*, pp. 109, 182–3, 194.

20. A. Hardy, 'Scientific Strategy and Ad Hoc Response: The Problem of Typhoid in America and England, *c.* 1910–50', *Journal of the History of Medicine and Allied Sciences* (April 2012), pp. 1–35, on pp. 27, 29; W. V. Shaw, 'Report on an Outbreak of Enteric Fever in the County Borough of Bournemouth and in the Boroughs of Poole and Christchurch', *Reports on Public Health and Medical Subjects*, 81, Ministry of Health (London: HMSO, 1937), p. 16.

21. Foster, *History of Medical Bacteriology*, pp. 191–4; J. Craigie and C. H. Yen, 'The Demonstration of Types of B. Typhosus by Means of Preparations Type II Vi Phage', *Canadian Public Health Journal*, 29 (1938), pp. 448–63, 484–96.

22. Ibid., pp. 484–91.
23. 'Serological Diagnosis of Typhoid', *British Medical Journal* (10 December 1938), pp. 1211–12.
24. Foster, *History of Medical Bacteriology*, p. 195.
25. Gradmann, *Laboratory Disease*, pp. 2–3, 6, 21, 53–4.
26. E. M. Crookshank, *A Text-Book of Bacteriology including the Etiology and Prevention of Infective Diseases*, 4th edn (London: H. K. Lewis, 1896), pp. 65–169.
27. C. Crenner, *Private Practice: In the Early Twentieth-Century Medical Office of Dr Richard Cabot* (Baltimore, MD: The Johns Hopkins University Press, 2005), pp. 39–41.
28. See Vernon, 'Pus, Sewage, Beer and Milk', p. 297, on side rooms.
29. See p. 58. T. J. Horder, *Clinical Pathology in Practice* (London: Henry Frowde, Hodder and Stoughton and Oxford University Press, 1910), p. v.
30. See Worboys, *Spreading Germs* and Vernon, 'Pus, Sewage, Beer and Milk', for detailed accounts of bacteriological research in Britain.
31. Ibid., pp. 295–9; Worboys, *Spreading Germs*, pp. 263–4.
32. Vernon, 'Pus, Sewage, Beer and Milk', pp. 297–8.
33. Worboys, *Spreading Germs*, pp. 187, 215, 232, 239–40, 264–5.
34. Vernon, 'Pus, Sewage, Beer and Milk', pp. 293, 303–5.
35. Worboys, *Spreading Germs*, p. 6.
36. By spring 1891 it was proven to be ineffective as a therapy, though it was later found to be useful as a diagnostic skin test. Gradmann, *Laboratory Disease*, p. 15.
37. Mendelsohn, 'Cultures of Bacteriology', pp. 442–6.
38. Ibid., p. 447; Worboys, *Spreading Germs*, pp. 212–13, 231.
39. Ibid., pp. 226–7, 256–7, 265.
40. Ibid., p. 261.
41. See J. H. Warner, *The Therapeutic Perspective: Medical Practice, Knowledge, and Identity in America, 1820–1885* (Princeton, NJ: Princeton University Press, 1997) and J. D. Howell, *Technology in the Hospital: Transforming Patient Care in the Early Twentieth Century* (Baltimore, MD, and London: The Johns Hopkins University Press, 1995) as examples of the significance of rigorous research with clinical case notes. See G. B. Risse and J. H. Warner, 'Reconstructing Clinical Activities: Patient Records in Medical History', *Social History of Medicine*, 5 (1992), pp. 183–205, on the value of case notes for historical research.
42. Christopher Lawrence has been particularly influential in creating the impression that the laboratory and specialization were not enthusiastically received by British gentlemen physicians: C. Lawrence, 'Incommunicable Knowledge: Science, Technology and the Clinical Art in Britain, 1850–1914', *Journal of Contemporary History*, 20 (1985), pp. 503–20; C. Lawrence, 'Still Incommunicable: Clinical Holists and Medical Knowledge in Interwar Britain', in C. Lawrence and G. Weisz (eds), *Greater Than The Parts: Holism in Biomedicine, 1920–1950* (New York and Oxford: Oxford University Press, 1998), pp. 94–111; C. Lawrence, 'A Tale of Two Sciences: Bedside and Bench in Twentieth-Century Britain', *Medical History*, 43 (1999), pp. 421–49; C. Lawrence, 'Edward Jenner's Jockey Boots and the Great Tradition in English Medicine 1918–1939', in C. Lawrence and A. K. Mayer (eds), *Regenerating England: Science, Medicine and Culture in Inter-War Britain* (Amsterdam: Rodopi, 2000), pp. 45–65. Examples of historical work influenced by the tensions genre include: L. S. Jacyna, 'The Laboratory and the Clinic: The Impact of Pathology on Surgical Diagnosis in the Glasgow Western Infirmary, 1875–1910', *Bulletin of the History of Medicine*, 62 (1988), pp. 384–406; A. Digby, *The Evolution of British General Practice 1850–1948* (Oxford: Oxford University Press, 1999), p. 222;

T. Alborn, 'Insurance against Germ Theory: Commerce and Conservatism in Late-Victorian Medicine', *Bulletin of the History of Medicine*, 75 (2001), pp. 406–45; M. Pelling, 'Contagion/Germ Theory/Specificity', in W. F. Bynum and R. Porter (eds), *Companion Encyclopaedia of the History of Medicine, Volume 1* (London and New York: Routledge, 1993), pp. 309–34, on p. 329. For an international context of this idea of tensions between the clinic and the laboratory, see R. C. Maulitz, '"Physician Versus Bacteriologist": The Ideology of Science in Clinical Medicine', in M. J. Vogel and C. E. Rosenberg (eds), *The Therapeutic Revolution: Essays in the Social History of American Medicine* (Philadelphia, PA: University of Pennsylvania Press, 1979), pp. 91–107; G. L. Geison, 'Divided We Stand: Physiologists and Clinicians in the American Context', in Vogel and Rosenberg (eds), *The Therapeutic Revolution*, pp. 67–90; J. H. Warner, 'The Fall and Rise of Professional Mystery: Epistemology, Authority and the Emergence of Laboratory Medicine in Nineteenth-Century America', in A. Cunningham and P. Williams (eds), *The Laboratory Revolution in Medicine* (Cambridge: Cambridge University Press, 2002), pp. 110–41; C. Timmermann, 'Constitutional Medicine, Neoromanticism and the Politics of Antimechanism in Interwar Germany', *Bulletin of the History of Medicine*, 75 (2001), pp. 717–39.

43. A. Cunningham and P. Williams, 'Introduction', in Cunningham and Williams (eds), *Laboratory Revolution*, pp. 1–13, on p. 11.

44. Howell, *Technology in the Hospital*, pp. 201–7; G. Davis, *'The Cruel Madness of Love': Sex, Syphilis and Psychiatry in Scotland, 1880–1930* (Amsterdam and New York: Rodopi, 2008), pp. 130–9; M. Hammerborg, 'The Laboratory and the Clinic Divide Revisited: The Introduction of Laboratory Medicine at the Bergen Hospital, Norway', *Social History of Medicine*, 24 (2011), pp. 758–75. C. Lawrence has examined a biochemical clinical laboratory in Edinburgh using laboratory reports and some case notes. He found that during the 1920s the bacteriological department was called upon for specialist forms of assistance. However, most diagnoses only used urine tests, symptoms and signs. C. Lawrence, *Rockefeller Money, The Laboratory, and Medicine in Edinburgh, 1919–1930: New Science in an Old Country* (Rochester, NY, and Woodbridge: University of Rochester and Boydell and Brewer, 2005), pp. 186–224.

45. C. Crenner, 'Professional Measurement: Quantifying Health and Disease in American Medical Practice, 1880–1920' (PhD dissertation, Harvard University, 1993), pp. 99–100; N. J. Tomes, 'American Attitudes toward the Germ Theory of Disease: Phyllis Allen Richmond Revisited', *Journal of the History of Medicine and Allied Sciences*, 52 (1997), pp. 17–50, on pp. 17–18; P. Palladino, 'On Writing the Histor(ies) of Modern Medicine', *Rethinking History*, 3 (1999), pp. 271–88; Crenner, *Private Practice*; A. J. Hull, 'Teamwork, Clinical Research, and the Development of Scientific Medicine in Interwar Britain: The Glasgow School Revisited', *Bulletin of the History of Medicine*, 81 (2007), pp. 569–93; S. Sturdy, 'Looking for Trouble: Medical Science and Clinical Practice in the Historiography of Modern Medicine', *Social History of Medicine*, 24 (2011), pp. 739–57.

46. These hospitals were identified using the Wellcome Trust and National Archives Hospital Records Database, at www.nationalarchives.gov.uk/hospitalrecords [accessed October to December 2003]. Thanks to Lesley Hall for her assistance with advanced searches of the database.

47. B. Abel-Smith, *The Hospitals 1800–1948: A Study in Social Administration in England and Wales* (London: Heinemann, 1964), pp. 38, 45, 126–7.

48. 'Report on the Sanitary Condition of the Borough of Cambridge, 1907', Cambridgeshire Collection, Milton Road Library, Cambridge, p. 21.

49. Statistical Tables of the Patients Under Treatment in the Wards of St. Bartholomew's Hospital, 1910–1920, St Bartholomew's Hospital Archives (hereafter SBHA).

50. See, for example Lawrence, 'Incommunicable Knowledge'; Lawrence, 'Still Incommunicable'; Lawrence, 'A Tale of Two Sciences'. For discourse in American hospitals, see Warner, *Therapeutic Perspective*, pp. 87–91.

51. G. L. Geison, *Michael Foster and the Cambridge School of Physiology: The Scientific Enterprise in Late Victorian Society* (Princeton, NJ: Princeton University Press, 1978); M. W. Weatherall, *Gentlemen, Scientists and Doctors: Medicine at Cambridge, 1880–1940* (Woodbridge and Rochester, NY: The Boydell Press in association with Cambridge University Library, 2000).

52. K. Waddington, *Medical Education at St. Bartholomew's Hospital 1123–1995* (Woodbridge: The Boydell Press, 2003), p. 75.

53. Ibid., pp. 128, 140, 147–8.

54. Crenner, 'Professional Measurement', p. 174.

55. Warner, *Therapeutic Perspective*, pp. 87–91.

56. Gradmann, *Laboratory Disease*, pp. 52–3.

57. Waddington, *The Bovine Scourge*, pp. 96–100.

58. N. Pemberton and M. Worboys, *Mad Dogs and Englishmen: Rabies in Britain, 1830–2000* (Basingstoke: Palgrave Macmillan, 2007), pp. 111–15.

59. Tanneries are very briefly mentioned by S. Jones, *Death in a Small Package: A Short History of Anthrax* (Baltimore, MD: The Johns Hopkins University Press, 2010), p. 114, and J. T. Carter, 'Anthrax in Kidderminster, 1900–1914' (PhD dissertation, University of Birmingham, 2005), p. 40; C. Holmes, *Spores, Plagues and History: The Story of Anthrax* (Dallas, TX, Durban House Pub., 2003), pp. 1, 23; R. M. Swiderski, *Anthrax: A History* (Jefferson, NC, and London: McFarland & Co. 2004), pp. 1, 32, 159; I. Mortimer and J. Melling, '"The Contest between Commerce and Trade, on the One Side, and Human Life on the Other": British Government Policies for the Regulation of Anthrax Infection and the Wool Textiles Industries, 1880–1939', *Textile History*, 31:2 (2000), p. 223.

60. 'Health International: Research on Anthrax', *The Times*, 14 September 1923, p. 7.

61. Worboys, *Spreading Germs*, p. 266; L. G. Stevenson, 'Exemplary Disease: The Typhoid Pattern', *Journal of the History of Medicine and Allied Sciences*, 37 (1982), pp. 159–81, on pp. 161–2.

62. J. F. Witt, *Lessons from History: State Constitutions, American Tort Law, and the Medical Malpractice Crisis*, The Project on Medical Liability in Pennsylvania funded by the Pew Charitable Trust, 2004, at http://www.pewtrusts.org/uploadedFiles/wwwpewtrustsorg/Reports/Medical_liability/medical_malpractice_witt_030904.pdf [accessed 18 May 2012].

63. Hardy, 'Scientific Strategy', p. 18.

1 Using Bacteriology in Teaching Hospitals: London and Cambridge, 1880–1920

1. 'Notes from the Wards', *St Bartholomew's Hospital Journal*, 3 (1895–6), p. 63.

2. See Sturdy, 'Looking for Trouble'; R. Wall, 'Using Bacteriology in Elite Hospital Practice: London and Cambridge, 1880–1920', *Social History of Medicine*, 24 (2011), pp. 776–95.

3. Waddington, *Medical Education*, pp. 1, 110.

4. The physicians of Bart's are particularly well represented in Christopher Lawrence's work as exemplars of the importance of character for the practice of medicine, including Patrick Black, Dyce Duckworth and Sir James Paget in Lawrence, 'Incommunicable Knowledge'. A high prominence is given to Thomas Horder in his second article on the subject, alongside Samuel Gee, Sir Percival Horton-Smith-Hartley, and Sir Walter Langdon Brown: Lawrence, 'Still Incommunicable'. Lawrence, 'A Tale of Two Sciences' is entirely focused on the Bart's physicians Gee, Horder and Brown. Another chapter looks at Gee, Horder, Brown, the surgeon D'Arcy Power, and even William Harvey, amongst other examples from the London and Oxford elites: Lawrence, 'Edward Jenner's Jockey Boots'. In Lawrence's first article on this topic, though, Thomas Lauder Brunton is portrayed as an advocate of diagnostic technologies such as the sphygmograph and sphygmomanometer, and fan of continental science, without mentioning he was a Bart's physician: see Lawrence, 'Incommunicable Knowledge', pp. 516–17. Tensions between the laboratory and elite physicians in the clinic have also been examined by others in the context of London and England, and in other countries, as mentioned in the introduction to this volume, p. 7.

5. Lawrence, 'Incommunicable Knowledge', pp. 503–4.

6. R. Porter, *Bodies Politic: Disease, Death and Doctors in Britain, 1650–1900* (Ithaca, NY: Cornell University Press, 2001), p. 255.

7. M. J. Peterson, *The Medical Profession in Mid-Victorian London* (Berkeley and Los Angeles, CA, and London: University of California Press, 1978), p. 137.

8. Ibid., p. 136.

9. P. Ferris, *The Doctors* (London: Victor Gollancz Ltd, 1965), p. 612.

10. Ibid., pp. 62, 64.

11. Lawrence, 'Incommunicable Knowledge', p. 507.

12. Ibid., pp. 515–16.

13. Lawrence, 'Still Incommunicable', pp. 94–5.

14. N. Moore, *The History of St. Bartholomew's Hospital, Volume II* (London: C. Arthur Pearson Limited, 1918); D. Power, *A Short History of St Bartholomew's Hospital, 1123–1923: Past and Present* (London: Saint Bartholomew's Hospital, 1923); G. Whitteridge and V. Stokes, *A Brief History of the Hospital of Saint Bartholomew* (London: The Governors of the Hospital of Saint Bartholomew, 1961); R. Bodley Scott, 'Medicine in the Twentieth Century', in V. C. Medvei and J. L. Thornton (eds), *The Royal Hospital of Saint Bartholomew, 1123–1973* (London: Saint Bartholomew's Hospital, 1974), pp. 185–204; Waddington, *Medical Education*, pp. 115–45, only presents this delay during the 1880s, not the 1890s.

15. A. Rook, M. Carlton and W. G. Cannon, *The History of Addenbrooke's Hospital, Cambridge* (Cambridge: Cambridge University Press, 1991), pp. 118–19.

16. Weatherall, *Gentlemen, Scientists and Doctors*, p. 15.

17. Ibid., p. 110; Rook et al., *Addenbrooke's*, pp. 118–19.

18. Weatherall, *Gentlemen, Scientists and Doctors*, p. 131.

19. Ibid., p. 116.

20. Lawrence, 'Incommunicable Knowledge', pp. 508–9, 511–15.

21. Weatherall, *Gentlemen, Scientists and Doctors*, p. 332.

22. Ibid.

23. Geison, *Michael Foster*, p. 108. The professorship was not created at this time due to agricultural depression lowering Trinity College's income.

24. Weatherall, *Gentlemen, Scientists and Doctors*, pp. 133, 139–40.

25. Ibid., p. 143. Weatherall cites *Lancet*, 1 (1884), p. 911.

26. Rook et al., *Addenbrooke's*, pp. 58, 238. The hospital was opened in 1766, with the Regius Professor of Physic, Russell Plumptre, as a physician to Addenbrooke's.

27. Ibid., pp. 60, 61, 83, 85, 101.

28. Ibid., p. 238, citing W. Langdon Brown, *Some Chapters in Cambridge Medical History* (1947); Geison, *Michael Foster*, pp. 365–6.

29. Weatherall, *Gentlemen, Scientists and Doctors*, p. 189.

30. Rook et al., *Addenbrooke's*, pp. 238–9.

31. Ibid., pp. 247–8, 315.

32. Geison, *Michael Foster*, pp. 9, 81, 96.

33. Ibid., p. 47; Weatherall, *Gentlemen, Scientists and Doctors*, pp. 94–8.

34. Weatherall, *Gentlemen, Scientists and Doctors*, p. 88; Geison, *Michael Foster*, p. 123.

35. Worboys, *Spreading Germs*, pp. 187, 264, 282; A. Hardy, 'On the Cusp: Epidemiology and Bacteriology at the Local Government Board, 1890–1905', *Medical History*, 42 (1998), pp. 328–46, on p. 341; Waddington, *Medical Education*, pp. 133–5.

36. Moore, *St. Bartholomew's Hospital, Volume II*, p. 754.

37. Waddington, *Medical Education*, pp. 110–17, 140–2.

38. Minute Book of Meetings of the Treasurer and Almoners and the Medical Council 1891–1902, Meeting, 17 January 1895, SBHA, Administration, Clerk's Office, Ha 6/5. Keir Waddington has also discussed this debate in less detail: Waddington, *Medical Education*, pp. 110–17, 140–2.

39. G. H. Brown (ed.), *Munk's Roll, Volume IV: Lives of the Fellows of the Royal College of Physicians of London, 1826–1925* (London: Royal College of Physicians, 1955), pp. 177–8.

40. Medical Registers, SBHA, MR 16/30, 16/36, 16/37, 16/42, 16/46 and 16/52. When St Bartholomew's case books or notes (Medical Registers, 1881–1920, SBHA, MR 16) are mentioned individually they are referenced SBHA, MR 16/number of case book/ number of case file.

41. Meeting, 17 January 1895, SBHA, Ha 6/5; Moore, *St. Bartholomew's Hospital, Volume II*, p. 801.

42. Dr Shore, Meeting, 17 January 1895, SBHA, Ha 6/5.

43. Meeting, 17 January 1895, SBHA, Ha 6/5. See also Moore, *St. Bartholomew's Hospital, Volume II*, p. 247.

44. F. Andrewes, 'The Beginnings of Bacteriology at Bart's, being Part of an Address to the Abernethian Society', *St Bartholomew's Hospital Journal* (May, 1928), pp. 116–17, on p. 116.

45. Minutes of the Board of Governors 11 May 1893–12 November 1903, House Committee Meeting, 14 March 1895, SBHA, Administration, Clerk's Office, Ha 1/27; Meeting, 15 May 1902, SBHA, Ha 6/5.

46. Andrewes, 'Beginnings of Bacteriology', p. 116.

47. Treasurer, Meeting, 15 May 1902, SBHA, Ha 6/5.

48. Letter from the Medical Council, signed by H. H. Tooth, Hon. Sec., 10 April 1902, read out by the Treasurer, Meeting, 15 May 1902, SBHA, Ha 6/5.

49. Meeting, 15 May 1902, SBHA, Ha 6/5.

50. House Committee Meeting, 14 March 1895, including a reading out of a letter from the Medical Council written by Anthony Bowlby, 14 February 1895, Minutes of the Board of Governors 11 May 1893–12 November 1903, SHBA, Administration, Clerk's Office, Ha 1/27.

51. Andrewes, 'Beginnings of Bacteriology', p. 116. See also Worboys, *Spreading Germs*, p. 292, for the career aspirations of most pathologists.
52. I. Waddington, *The Medical Profession in the Industrial Revolution* (Dublin: Gill and Macmillan Ltd, 1984), p. 32.
53. Treasurer, Meeting, 17 January 1895, SBHA, Ha 6/5.
54. Meeting, 15 May 1902, SBHA, Ha 6/5.
55. Meeting, 10 July 1902, SBHA, Ha 6/5.
56. Treasurer and Almoners' Minute Book May 1903 to December 1904, Meetings, 23 June 1903 and 15 December 1904, SBHA, Ha 3/26.
57. Treasurer and Almoners' Minute Book 1905 to 1906, Meeting, 9 March 1905, SBHA, Ha 3/27.
58. Medical Council Minutes, Special Meeting, 31 January 1906, SBHA, MC 1/3.
59. Treasurer's Reports, 1908, SBHA, Ha 103/2, p. 7.
60. Treasurer's Reports, 1908, SBHA, Ha 103/2, pp. 15–16.
61. Ibid.
62. Moore, *St. Bartholomew's Hospital, Volume II*, pp. 754–5.
63. F. Andrewes, 'The Meaning of the New Pathological Block', *St Bartholomew's Hospital Journal* (June 1909), p. 138.
64. Cited in Waddington, *Medical Education*, pp. 154–5.
65. Weatherall, *Gentlemen, Scientists and Doctors*, pp. 144, 147; Meeting of the General Committee, 6 July 1908, 'The Pathological Department', insert, Addenbrooke's Hospital Minutes, Addenbrooke's Hospital Archives (hereafter AHA), AHM 45, p. 318; Rook et al., *Addenbrooke's*, p. 256.
66. Meeting, 6 March 1895, AHA, AHM 35, p. 531; 20 November 1895, AHA, AHM 35, pp. 189–90.
67. Rook et al., *Addenbrooke's*, pp. 256–7.
68. Committee Meeting of Repairs Committee, 25 October 1899, AHA, AHM 39, p. 149.
69. Rook et al., *Addenbrooke's*, pp. 256–7.
70. Weatherall, *Gentlemen, Scientists and Doctors*, p. 154. Meeting of the General Committee, 6 July 1908, 'The Pathological Department', insert, AHA, AHM 45, 1907–9, p. 318. Rook et al., *Addenbrooke's*, p. 257; Worboys, *Spreading Germs*, p. 216.
71. Meeting of the General Committee, 6 July 1908, 'The Pathological Department', insert, AHA, AHM 45, 1907–9, p. 318.
72. Rook et al., *Addenbrooke's*, pp. 212, 242, 257, 259.
73. Ibid., p. 242.
74. E. H. Ackerknecht, 'A Plea for a "Behaviourist" Approach in Writing the History of Medicine', *Journal of the History of Medicine and Allied Sciences*, 22 (1967), pp. 211–14, on pp. 211, 214.
75. Risse and Warner, 'Reconstructing Clinical Activities'.
76. Case books are the main source for this chapter and the first half of Chapter 2, hence they are not referenced each time they are analysed as a group in the text or in figures and tables. As mentioned in note 40 above, when St Bartholomew's case books or notes (Medical Registers, 1881–1920, SBHA, MR 16) are mentioned individually they are referenced SBHA, MR 16/number of case book/number of case file. When Addenbrooke's Hospital case notes (medical casebooks, 1880–1920, AHA, AHPR1/1) are mentioned individually they are referenced AHA, AHPR1/1 box number/book number or reference/number of case file.

77. G. Bourne, *We Met at Bart's: The Autobiography of a Physician* (London: Friedrich Muller Limited, 1963), pp. 16, 35–7, 78, 123.

78. SBHA, MR 16/39, 16/43.

79. SBHA, MR 16/64/190; SBHA, MR 16/64/191; SBHA, MR 16/69/63.

80. I have become very familiar with Samuel Gee's handwriting from looking through his private case books which are discussed in Chapter 2, pp. 52–3. I have not attempted this kind of analysis for any other physician as I do not have enough familiarity with their handwriting.

81. Rook et al., *Addenbrooke's*, pp. 179–80; Addenbrooke's Hospital Annual Reports, 1880–1920, AHA.

82. Rook et al., *Addenbrooke's*, pp. 179–80.

83. See R. Wall, 'Using Bacteriology in the Hospital and Society: England 1880–1939' (PhD dissertation, University of London, 2007), appendices, for a breakdown of the number of cases per year.

84. See p. 8.

85. See discussions in Meetings, 1 November to 15 November 1899, AHA, AHM 39, pp. 159–68.

86. K. Waddington, *Charity and the London Hospitals, 1850–1898* (Woodbridge: Boydell and Brewer, 2000), p. 12.

87. Weekly Meeting, 7 November 1900, AHA, AHM 40, pp. 57–8.

88. Weekly Meeting, 18 March 1903, AHA, AHM 42, pp. 40–1.

89. Ibid.; Meetings of the House Sub-Committee, 14 November 1904, p. 193; 28 November 1904, pp. 208–67; 10 December 1904, AHA, AHM 43, pp. 226–7.

90. Special Meeting of the House Committee, 4 March 1905, AHA, AHM 43, pp. 368–9.

91. See 'Report of the Sanitary Condition of the Cambridge Improvement Act District', 1880–8; 'Report of the Sanitary Condition of the Borough of Cambridge', 1889–1900, 1902–20, Cambridgeshire Collection, Milton Road Library, Cambridge.

92. Weatherall, *Gentlemen, Scientists and Doctors*, pp. 155–6; Rook et al., *Addenbrooke's*, p. 257; Borough of Cambridge, Public Health Committee Minute Book, 22 June 1899 to 21 April 1904, Meetings of the Public Health Committee, 22 September 1899, p. 16; 3 October 1899, p. 26; 29 October 29 1899, p. 113; 10 December 1901, pp. 210–11; 21 January 1902, Cambridgeshire County Record Office, Cambridge (hereafter CCRO), p. 222.

93. Borough of Cambridge, Public Health Committee Minute Book, 22 June 1899 to 21 April 1904, Meetings of the Public Health Committee, 10 December 1901, pp. 210–11; 21 January 1902, CCRO, p. 222.

94. A study of this kind would be difficult, as it seems routine tests were often carried out. For an example, see the assessment of Thomas Horder and the Wasserman test for syphilis in Chapter 2, p. 58.

95. Lawrence, 'Incommunicable Knowledge', p. 508.

96. Howell, *Technology in the Hospital*, pp. 81–2, 92.

97. F. Condrau and M. Worboys, 'Second Opinions: Epidemics and Infections in Nineteenth-Century Britain', *Social History of Medicine*, 20 (2007), pp. 147–58, on p. 153.

98. See p. 4.

99. E. M. Hammonds, *Childhood's Deadly Scourge: The Campaign to Control Diphtheria in New York City* (Baltimore, MD, and London: The John Hopkins University Press, 1999), p. 50; Worboys, *Spreading Germs*, p. 193.

100. See p. 52.

101. Jacyna, 'The Laboratory and the Clinic', esp. pp. 386–7; Davis, *The Cruel Madness of Love*. Worboys has also discussed the problems of false negative results from bacteriological tests for tuberculosis, and patients who appeared healthy, but had tubercle bacilli in their sputa. See Worboys, *Spreading Germs*, pp. 213, 215.
102. AHA, AHPR 1/1/006/1-1-8/141.
103. SBHA, MR 16/40/158.
104. B. Pasveer, 'Depiction in Medicine as a Two-Way Affair: X-Ray Pictures and Pulmonary Tuberculosis in the Early Twentieth Century', in I. Löwy (ed.), *Medicine and Change: Historical and Sociological Studies of Medical Innovation. Proceedings of the Symposium INSERM held in Paris, 21–23 April, 1992* (London and Paris: John Libbey Eurotext and INSERM, 1993), p. 87.
105. C. L. Leonard, 'Discussion on the Diagnosis of Pulmonary Tuberculosis by Means of the Roentgen Rays', *British Medical Journal*, 2 (1908), p. 706; R. M. Leslie, 'Hilus Tuberculosis ("root phthisis")', *British Journal of Tuberculosis*, 7 (1913), pp. 163–70, both cited in Pasveer, 'Depiction in Medicine', pp. 88–9. Also see ibid., pp. 90–1 for doubts in the 1920s.
106. Lindqvist, 'Change in the Technological Landscape', pp. 271–88.
107. A. Hardy, 'Tracheotomy versus Intubation: Surgical Intervention in Diphtheria in Europe and the United States, 1825–1930', *Bulletin of the History of Medicine*, 66 (1992), p. 536–59, on p. 536; Hammonds, *Childhood's Deadly Scourge*, p. 19.
108. Hardy, 'Tracheotomy versus Intubation', pp. 536, 542.
109. C. Hooker and A. Bashford, 'Diphtheria and Australian Public Health: Bacteriology and its Complex Applications, c. 1890–1930', *Medical History*, 46 (2002), pp. 41–64, on p. 46.
110. P. Weindling, 'From Medical Research to Clinical Practice: Serum Therapy for Diphtheria in the 1890s', in J. V. Pickstone (ed.), *Medical Innovations in Historical Perspective* (New York: St. Martin's Press, 1992), pp. 72–83, on p. 73; Hammonds, *Childhood's Deadly Scourge*, pp. 8, 17, 19.
111. Hooker and Bashford, 'Diphtheria and Australian Public Health', p. 48.
112. Hammonds, *Childhood's Deadly Scourge*, pp. 50–2.
113. Andrewes, 'Beginnings of Bacteriology', p. 116.
114. Statistical Tables of the Patients under Treatment in the Wards of St. Bartholomew's Hospital, 1896, SBHA.
115. In one anomalous year, 1920, the bacteriological test was recorded in fewer case notes than the membrane test.
116. AHA, AHPR 1/1/031/1-1-70/044.
117. Weindling, 'From Medical Research to Clinical Practice', pp. 75–7.
118. Dr. Shore, Meeting, 17 January 1895, SBHA, Ha 6/5.
119. Hardy, 'Tracheotomy versus Intubation', p. 547.
120. Ibid., p. 550.
121. Ibid., pp. 548–50.
122. SBHA, MR 16/42/67.
123. See p. 14, n. 4 for how Lauder Brunton has been portrayed by Lawrence.
124. C. W. LeBaron and D. N. Taylor, 'Typhoid Fever', in K. F. Kiple (ed.), *The Cambridge World History of Human Disease* (Cambridge: Cambridge University Press, 1999), pp. 1071–7, on p. 1075.
125. Howell, *Technology in the Hospital*, p. 203.
126. Ibid., p. 204.
127. See for example SBHA, MR 16/57/21 and AHA, AHPR 1/1/021/PB/072.

128. Waddington, *Charity*, p. 12.
129. The physicians whose pulmonary tuberculosis, diphtheria and typhoid case notes have uniquely survived for these thirteen years are: 1881 – Dr James Matthews Duncan; 1882 – Dr Samuel Gee; 1894 – Dr Samuel Gee; 1897 – Sir Dyce Duckworth; 1898 – Sir William Church; 1905 – Sir Wilmot Herringham; 1908 – Sir Norman Moore; 1910 – Dr Howard Tooth; 1912 – Dr Howard Tooth; 1913 – Dr Howard Tooth; 1916 – Dr Herbert Morley Fletcher; 1917 – Dr John Drysdale; 1919 – Dr Howard Tooth.

2 Integrating the Laboratory into Gentlemanly Practice

1. J. W. Legg, 'Recollections of Samuel Gee, Physician to Saint Bartholomew's Hospital brought together by J. W. Legg' (1911), in S. Gee, *Medical Lectures and Aphorisms*, with recollections by J. W. Legg (London: Henry Frowde, Oxford University Press and Hodder and Stoughton, 1902, 1915), pp. 353–92, on pp. 379–80.
2. As in Chapter 1, case books are a main source for this chapter. They are not referenced each time they are analysed as a group. When St Bartholomew's case books or notes are mentioned individually they are referenced SBH 16/Number of case book/Number of case file. When Addenbrooke's Hospital case notes are mentioned individually they are referenced AH Box number/Book number or reference/Number of case file. See p. 192, n. 76.
3. Brown, *Munk's Roll*, p. 183. Gee's last case notes to survive are from 1899. See also A. White Franklin, 'Medical Achievements of the Eighteenth and Nineteenth Centuries', in Medvei and Thornton (eds), *The Royal Hospital of Saint Bartholomew 1123–1973*, pp. 126–84, on pp. 177–8.
4. R. R. Trail (ed.), *Munk's Roll, Volume V: Lives of the Fellows of the Royal College of Physicians of London, continued to 1965* (London: Royal College of Physicians, 1968), p. 198.
5. W. Ernst, 'The Normal and the Abnormal: Reflections on Norms and Normativity', in W. Ernst, *Histories of the Normal and the Abnormal: Social and Cultural Histories of Norms and Normativity* (Abingdon and New York: Routledge, 2006), pp. 1–25, on p. 4.
6. Warner, *Therapeutic Perspective*, pp. 85–6.
7. Ernst, 'The Normal and the Abnormal', p. 4.
8. Ibid., p. 2.
9. I. Hacking, *The Taming of a Chance* (Cambridge: Cambridge University Press, 1990), p. 169, cited in Ernst, 'The Normal and the Abnormal', p. 3. For critique of the term, see Ernst's analysis of literature by Michel Foucault and Georges Canguilhem, ibid., pp. 6–7. C. Sinding, 'The Power of Norms: Georges Canguilhem, Michel Foucault, and the History of Medicine', in F. Huisman and J. H. Warner (eds), *Locating Medical History: The Stories and their Meanings* (Baltimore, MD, and London: The Johns Hopkins University Press, 2004), pp. 262–84, on pp. 272–3.
10. Warner, *Therapeutic Perspective*, pp. 86–7.
11. Terrie Romano frames physiology in Britain as the study of the 'normal'. See T. M. Romano, *Making Medicine Scientific: John Burdon Sanderson and the Culture of Victorian Science* (Baltimore, MD, and London: The Johns Hopkins University Press, 2002), p. 2.
12. See C. Bernard, 'An Introduction to the Study of Experimental Medicine' (1865), in C. Bernard, *Experimental Medicine* with a new introduction by S. Wolf, trans. H. C. Greene (New Brunswick, NJ, and London: Transaction Publishers, 1999), pp. 2, 71, 84, 149.
13. G. Canguilhem, *The Normal and the Pathological* (New York: Zone Books, 1991), pp. 88–9. Also see Ernst, 'The Normal and the Abnormal', pp. 6–7.
14. Ibid., p. 3.

15. A. Matthews David, 'Made to Measure? Tailoring and the "Normal" Body in Nine-teenth-Century France', in Ernst, *Histories of the Normal and the Abnormal*, pp. 142–64, on p. 143.

16. C. Sinding, 'Flexible Norms? From Patients' Values to Physicians' Standards', in Ernst, *Histories of the Normal and the Abnormal*, pp. 225–44, on p. 226; Sinding, 'The Power of Norms', pp. 267–70, 276.

17. Warner, *Therapeutic Perspective*, pp. 83–161.

18. Ibid., pp. 88, 104–13.

19. Ibid., p. 91.

20. Lawrence, 'Still Incommunicable', pp. 91, 97–8.

21. AHA, AHPR 025/8-9/208.

22. SBHA, MR 16/042/028.

23. Warner, *Therapeutic Perspective*, pp. 89–90.

24. Geison, *Michael Foster*, pp. 183–4.

25. Warner, *Therapeutic Perspective*, pp. 5–6, 85–7.

26. Worboys, *Spreading Germs*, pp. 31–2.

27. Warner, *Therapeutic Perspective*, p. 87.

28. P. F. Cranefield, 'The Organic Physics of 1847 and the Biophysics of Today', *Journal of the History of Medicine and Allied Sciences*, 12 (1957), pp. 407–23, on p. 407; R. G. Frank Jr, 'The Telltale Heart: Physiological Instruments, Graphic Methods, and Clinical Hopes, 1854–1914', in W. Coleman and F. L. Holmes (eds), *The Investigative Enterprise: Experimental Physiology in Nineteenth-Century Medicine* (Berkeley and Los Angeles, CA, and London, University of California Press, 1988), pp. 211–90, on pp. 215–16.

29. Geison, *Michael Foster*, pp. 106, 180–3, 253–6, 295.

30. Ibid., pp. 306, 308, 365–6.

31. Ibid., pp. 56, 330.

32. Ibid., pp. 150–1, 329.

33. White Franklin, 'Medical Achievements', pp. 180–3; Lawrence, 'Incommunicable Knowledge', pp. 516–17.

34. Waddington, *Medical Education*, pp. 125–6; Geison, *Michael Foster*, pp. 171–2.

35. Brown, *Munk's Roll*, pp. 183–4.

36. S. Gee, 'Notebook I, 1873–. Notes re Diseases, and Case-Notes', Library of the Royal College of Physicians, London (hereafter LRCPL), MS35, p. 233. See also S. Gee, 'Volume III of Notes, 1865–', LRCPL, MS34, p. 115, for a similar comment. Also see, Lawrence, 'Incommunicable Knowledge'.

37. Quoted in Bodley Scott, 'Medicine in the Twentieth Century', p. 185, Legg explains this decision. Apparently, Gee doubted experimental science, as there was doubt about the geometrical laws of triangles, and about the laws of gravity at this time. See Legg, 'Recollections', p. 363.

38. Quoted in Bodley Scott, 'Medicine in the Twentieth Century', p. 186. See a similar comment in his private notebook, S. Gee, 'Notebook V, 1879–. Notes re Diseases, and Case-Notes', LRCPL, MS36, p. 345.

39. Gee, LRCPL, MS36, pp. 135–6.

40. Ibid., p. 136.

41. Ibid., p. 305.

42. Ibid., p. 306.

43. Ibid., p. 259.

44. Andrewes, 'Beginnings of Bacteriology', p. 116.

45. S. Gee, 'On the Causes and Forms of Bronchitis', Lecture given before the Royal College of Physicians, in Gee, *Medical Lectures and Aphorisms*, pp. 60–75, on p. 74.

46. Ibid., p. 91. Lawrence quotes this statement too, modifying his view of Gee to an extent, but not showing the number of times, and the strength of Gee's arguments for the use of bacteriology. Lawrence, 'A Tale of Two Sciences', p. 427.

47. S. Gee, 'Clinical Aphorisms collected by Dr. Thomas Horder', in Gee, *Medical Lectures and Aphorisms*, pp. 243–304, on pp. 244, 266.

48. Ibid., p. 246.

49. Ibid., pp. 290–1.

50. S. Gee, 'Sects in Medicine', Read before the Abernethian Society on June 20, 1889, in Gee, *Medical Lectures and Aphorisms*, pp. 216–42, on p. 238.

51. Gee, LRCPL, MS34, pp. 123–4 – 'Medicine is the applicn of knowledge to the relief of sick men, it promises, as Celsus says, health to the sick. All branches of Knowledge are welcome which afford any help this way, anatomy, chemistry, pathology, therapeutics, electricity, cookery, upholstery.'

52. S. Gee, 'On the Nature of Pulmonary Emphysema', Lecture given before the Royal College of Physicians, in Gee, *Medical Lectures and Aphorisms*, pp. 95–128, on p. 95.

53. Legg, 'Recollections', p. 359.

54. Gee, 'Sects in Medicine', p. 216.

55. Ibid., p. 229.

56. Gee, LRCPL, MS35, p. 241.

57. Gee, 'Sects in Medicine', p. 236.

58. Ibid., pp. 239–40. See Lawrence, 'Edward Jenner's Jockey Boots', pp. 47–9, where Gee was represented as 'revering' empiricism.

59. Legg, 'Recollections', p. 363.

60. Gee, 'Sects in Medicine', p. 242.

61. Legg, 'Recollections', p. 380.

62. Gee, LRCPL, MS34, p. 124.

63. Ibid., pp. 126, 140–1.

64. Legg, 'Recollections', p. 363.

65. Gee, LRCPL, MS34, p. 169.

66. For the variety of roles, see M. W. Dupree, 'Other Than Healing: Medical Practitioners and the Business of Life Assurance during the Nineteenth and Early Twentieth Centuries', *Social History of Medicine*, 10 (1997) pp. 79–103, on p. 81.

67. M. Worboys, 'Private Clinical Laboratories in Britain: The Clinical Research Association, 1894–1914', unpublished paper presented at the American Association for the History of Medicine Annual Conference, Madison, 2004.

68. St Bartholomew's Hospital, 'Pathological Department of the Journal', p. 80.

69. Waddington, *Medical Education*, pp. 140, 143, 156.

70. H. Dale, 'Scientific Method in Medical Research: An Address Given on October 10, 1950 in Opening a Course of Lectures on "The Scientific Basis of Medicine" Arranged by the British Postgraduate Medical Federation', *British Medical Journal* (1950), pp. 1185–90, esp. p. 1187; Lawrence, 'Still Incommunicable', p. 96; K. D. Keele, *The Evolution of Clinical Methods in Medicine being the FitzPatrick Lectures delivered at the Royal College Physicians in 1960–61* (London: Pitman Medical Publishing Co. Ltd, 1963), p. 104.

71. Dale, 'Scientific Method', p. 1187.

72. E. M. Tansey, 'The Early Scientific Career of Sir Henry Dale FRS (1875–1968)' (PhD dissertation, University of London, 1990), pp. 75–98.

73. Ibid., pp. 2, 175–91.
74. D. Cantor, 'The MRC's Support for Experimental Radiology during the Inter-War Years', in J. Austoker and L. Bryder (eds), *Historical Perspectives on the Role of the MRC* (Oxford: Oxford University Press, 1989) pp. 181–204, on pp. 193–6.
75. See Lord Thomas Jeeves Horder, Lord Horder of Ashford Collection, Letter from Louise, Princess Royal, *c.* 1928–1929, Wellcome Library Manuscripts and Archives, GP/31/A.1/2; and Letter from Sybil Corkraw on behalf of Princess Beatrice (no date), Wellcome Library Manuscripts and Archives, GP/31/A.1/3.
76. Trail, *Munk's Roll*, pp. 198–200; M. Horder, *The Little Genius: A Memoir of the First Lord Horder* (London: Gerald Duckworth & Co Ltd, 1966), pp. 78–9.
77. Trail, *Munk's Roll*, pp. 198–200; E. Barnes, 'Fashioning a Natural Self: Guides to Self-Fashioning in Victorian England' (PhD dissertation, University of Cambridge, 1996), ch. 4.
78. See Lawrence, 'Still Incommunicable', p. 98, and 'A Tale of Two Sciences', p. 430.
79. Ibid., p. 431.
80. Andrewes, 'Beginnings of Bacteriology', p. 116.
81. T. Horder, 'Clinical Medicine as an Aid to Pathology: A Criticism, A Paper Read before the Abernethian Society, December, 1909', *St Bartholomew's Hospital Journal* (August, 1912), pp. 192–5, on p. 192.
82. Ibid., p. 192.
83. Ibid.
84. Ibid.
85. Horder, *Clinical Pathology in Practice*, p. v.
86. Ibid., pp. v, 2.
87. T. Horder, 'Medicine as a Career: An Opening Address given to the Welsh National Medical School on Oct. 6th, 1933', *Lancet*, 2 (1933), pp. 886–9, on p. 889.
88. T. Horder, 'On the Need for Standardisation in Clinical Pathology: Opening Address to the Association of British Pathologists', *Lancet*, 2 (1928), pp. 136–9, on p. 137. Lawrence also uses this quote, but omits Horder's fortune of having the services of the Bart's laboratory. See Lawrence, 'A Tale of Two Sciences', p. 445. Also see Lawrence, 'Still Incommunicable', pp. 105–6.
89. SBHA 16/81/2334.
90. T. Horder, 'The Strain of Modern Civilization: An Address to the British Association at Blackpool, 1936', in T. J. Horder, *Health and a Day: Addresses by Lord Horder* (London: J. M. Dent and Sons Ltd, 1937), pp. 1–18, on p. 2. He also mentions the 'exploitation of instruments of precision' in T. Horder, 'The Doctor as Humanist: An address at the Opening of the Westminster Hospital Medical School, 28th September 1936', in Horder, *Health and a Day*, pp. 19–32, on p. 32.
91. T. J. Horder, *Medical Notes* (London: Henry Frowde and Hodder and Stoughton, 1921), pp. 48–9. Emphasis his.
92. Horder, 'On the Need for Standardisation', p. 137.
93. T. Horder, 'The Clinician's Function in Medicine. Read at the Annual Meeting of the Medical Society of the State of New York, 28th April 1936', in Horder, *Health and a Day*, pp. 164–84, on pp. 171–2; T. Horder, 'Clinical Medicine: A Farewell Lecture', *Lancet*, 1(1936), pp. 179–82, on pp. 180–1.
94. S. A. Pai, 'Laboratory Tests: Proper Communication Reduces Error', *Student BMJ* (2005), p. 399.
95. See Lawrence, 'Still Incommunicable', p. 98, and Lawrence, 'A Tale of Two Sciences', pp. 430–1.

96. Romano, *Making Medicine Scientific*, pp. 3–4; P. Mandler, *The Fall and Rise of the English Stately Home* (New Haven, CT, and London: Yale University Press, 1997), p. 114.

97. D. Edgerton, *Science, Technology and the British Industrial 'Decline' 1870–1970* (Cambridge: Cambridge University Press, 1996), pp. 20, 67–9.

98. M. J. Wiener, *English Culture and the Decline of the Industrial Spirit, 1850–1980* (Cambridge: Cambridge University Press, 1981), pp. 17, 23.

99. E. D. Laborde, *Harrow School, Yesterday and Today* (London: Winchester Publications Limited, 1948), pp. 48, 55–6; 62; J. B. Hope Simpson, *Rugby Since Arnold: A History of Rugby School from 1842* (London and New York: Macmillan and St. Martin's Press, 1967), pp. 25–6. The Bart's physicians William Church, Joseph Ormerod and Howard Tooth attended either Harrow or Rugby.

100. Brown, *Munk's Roll*, pp. 177–8, 331, 304.

101. Hope Simpson, *Rugby Since Arnold*, pp. 25–6.

102. L. Daston and H. O. Sibum, 'Introduction: Scientific Personae and their Histories', *Science in Context*, 16 (2003), pp. 1–8.

103. E. Goffman, *The Presentation of Self in Everyday Life* (London: Allen Lane The Penguin Press, 1969).

104. Lawrence comments on diagnosis as the 'cornerstone of medicine', 'A Tale of Two Sciences', p. 427, repeated in 'Edward Jenner's Jockey Boots', p. 49. Hammonds also considers that change in diagnostic practices of diphtheria was seen as a threat to the medical profession of New York, *Childhood's Deadly Scourge*, p. 11.

105. Gee, 'Sects in Medicine', pp. 238–9.

106. Lawrence, 'A Tale of Two Sciences', p. 427.

107. T. J. Horder and A. E. Gow, *The Essentials of Medical Diagnosis: A Manual for Students and Practitioners*, rev. with the assistance of R. Bodley Scott (London, Toronto, Melbourne and Sydney: Cassell and Company Ltd, 1952), p. xiii.

108. Maulitz, "Physician Versus Bacteriologist", p. 92.

109. Gee, LRCPL, MS34, p. 137.

110. Howell, *Technology in the Hospital*, pp. 135–6, 159–60.

111. Ibid., p. 160. Howell cites S. Lange, 'The Present Status of the Roentgen Ray', *Lancet-Clinic*, 58 (1907), pp. 79–89; and E. H. Skinner, 'The Ownership of X-Ray Plates: Patient and Medical Attendant Entitled to Radiologist's Opinion, Not the Plate', *Modern Hospital*, 1 (1913), pp. 30–1.

112. Howell, *Technology in the Hospital*, p. 161.

113. Horder, 'The Doctor as Humanist', p. 32; Horder, 'The Clinician's Function in Medicine', pp. 174–5, 181.

114. T. Horder, 'Individuality in Medicine: An Address Delivered to the Students of Durham University Medical College at the Opening of the Winter Session, October 6th 1924', *Lancet*, 2 (1924), pp. 819–24, on p. 822.

115. Horder, 'The Clinician's Function in Medicine', p. 183.

116. Ibid., p. 174.

117. Ibid., p. 184.

118. T. Horder, 'Introduction', in Horder, *Health and a Day*, pp. vii–viii.

119. N. D. Jewson, 'The Disappearance of the Sick-Man from Medical Cosmology', *Sociology*, 10 (1976), pp. 225–44.

120. C. Crenner, 'Diagnosis and Authority in the Early Twentieth-Century Medical Practice of Richard C. Cabot', *Bulletin of the History of Medicine*, 76 (2002), pp. 30–55, esp. pp. 41–2, 45.

121. Digby, *British General Practice*, p. 287.

122. Ibid., p. 291.

123. Political and Economic Planning, *Report on the British Health Services: A Survey of the Existing Health Services in Great Britain with Proposals for Future Development, December 1937* (London: PEP, 1937), see for example, pp. 10–11 and 161.

124. Ibid., pp. 10–11, 161.

125. Ibid., see for example p. 11.

126. Ibid., pp. 160–1.

127. Ibid., p. 163.

128. Ibid., p. 164.

129. Digby, *British General Practice*, pp. 299–302.

130. A. Digby and N. Bosanquet, 'Doctors and Patients in an Era of National Health Insurance and Private Practice, 1913–1938', *Economic History Review*, new series, 41 (1988), pp. 74–94, esp. p. 92.

131. Horder, 'The Strain of Modern Civilization', p. 2. He mentions again the 'exploitation of instruments of precision', in Horder, 'The Doctor as Humanist', p. 32, and in Horder, 'The Clinician's Function in Medicine', p. 48.

132. See, for example, Horder, 'The Doctor as Humanist'.

133. Horder, 'The Strain of Modern Civilization', esp. pp. 12, 17.

134. Risse and Warner, 'Reconstructing Clinical Activities', p. 200.

3 Anthrax in Bradford: Understanding Deadly Disease in the Workplace, 1880–1905

1.   Bradford Medico-Chirurgical Society Minute Book, 1874–84, 4 April 1882, West Yorkshire Archive Service, Bradford (hereafter WYASB), 40D89/2.

2.   M. Bligh, *Dr. Eurich of Bradford* (London: James Clarke & Co. Ltd, 1960), p. 109, states 1847; N. Metcalfe, 'The History of Woolsorters' Disease: A Yorkshire Beginning with an International Future?', *Occupational Medicine*, 54 (2004), pp. 489–93, on p. 490, states 1838. Photocopy of newspaper cuttings regarding anthrax made by Dr. J. H. Bell until his death in 1906. A few later ones preserved by Dr F. W. Eurich, University of Bradford Special Collections, A15 (hereafter, Bell Newspaper cuttings).

3.   T. Carter and J. Melling, 'Trade, Spores, and the Culture of Disease: Attempts to Regulate Anthrax in Britain in its International Trade, 1875–1930', in C. Sellers and J. Melling, *Dangerous Trade: Histories of Industrial Hazard Across a Globalizing World* (Philadelphia, PA: Temple University Press, 2012), pp. 60–72, on p. 63; see also J. Stark, 'Industrial Illness in Cultural Context: La Maladie de Bradford in Local, National and Global Settings, 1878–1919' (PhD dissertation, University of Leeds, 2011), p. 210.

4.   P. Bartrip, *The Home Office and the Dangerous Trades: Regulating Occupational Disease in Victorian and Edwardian Britain* (Amsterdam: Rodopi, 2002), p. 254.

5.   L. Wilkinson, 'Anthrax', in K. F. Kiple (ed.), *The Cambridge World History of Human Disease* (Cambridge: Cambridge University Press, 1993), pp. 582–4, on p. 583; C. Alvin, 'Medical Treatment and Care in Nineteenth-Century Bradford: An Examination of Voluntary, Statutory, and Private Medical Provision in a Nineteenth-Century Urban Industrial Community' (PhD dissertation, University of Bradford, 1998), p. 277.

6.   Jones, *Death in a Small Package*, pp. xviii–6.

7. For example, Worboys, *Spreading Germs*, p. 43; S. Jones and P. Teigen, 'Anthrax in Transit: Practical Experience and Intellectual Exchange', *Isis*, 99 (2008), pp. 455–85, on p. 482.

8. Jones, *Death in a Small Package*, pp. 57–60.

9. Bartrip, *The Home Office*, p. 234; Jones, *Death in a Small Package*, pp. 51–5.

10. Vernon, 'Pus, Sewage, Beer and Milk', pp. 294–5.

11. Bartrip, *The Home Office*, p. 239.

12. Mortimer and Melling, "The Contest between Commerce and Trade", pp. 223, 231.

13. Wilkinson, 'Anthrax', p. 584.

14. Mortimer and Melling, "The Contest Between Commerce and Trade", pp. 226–7; Bartrip, *The Home Office*, p. 246.

15. J. Stark, 'Bacteriology in the Service of Sanitation: The Factory Environment and the Regulation of Industrial Anthrax in Late-Victorian Britain', *Social History of Medicine*, 25 (2012), pp. 343–61.

16. Bartrip, *The Home Office*, pp. 256–7.

17. Mortimer and Melling, "The Contest between Commerce and Trade".

18. J. Melling, 'Beyond a Shadow of a Doubt? Experts, Lay Knowledge, and the Role of Radiography in the Diagnosis of Silicosis in Britain, *c.* 1919–1945', *Bulletin of the History of Medicine*, 84 (2010), pp. 424–66, on p. 430.

19. Ibid., pp. 430–1.

20. Since completion of my doctoral research, Wall, 'Using Bacteriology' in 2007, James Stark has completed a doctoral thesis on anthrax in Bradford, 'Industrial Illness', see n. 3 above. Some of the research within the present chapter has been furthered within Stark's full-length study on the topic. Stark's thesis will be referred to when it helps to develop my arguments.

21. Jones, *Death in a Small Package*, p. 118.

22. Stark, 'Industrial Illness', p. 193.

23. Jones, *Death in a Small Package*, p. 117; Stark, 'Industrial Illness', pp. 198–9.

24. Bligh, *Dr. Eurich*; Mortimer and Melling, "The Contest between Commerce and Trade".

25. Short articles include Metcalfe, 'The History of Woolsorters' Disease'; A. Nicoll and R. Maynard, 'One Hundred Years of Anthrax', *Occupational and Environmental Medicine*, 61 (2004), p. 95. Other lengthier studies have various aims. Margaret Bligh's much-cited study of *Dr. Eurich of Bradford* is a eulogy about her father, and the chapters on anthrax are aimed at showing the importance of Drs Bell and Eurich in the fight against the disease in Bradford, and the resulting legislation. Mortimer and Melling, "The Contest between Commerce and Trade", examines anthrax in Bradford within a wider study of regulations and legislation regarding anthrax between 1880 and 1939. Carter and Melling, 'Trade, Spores, and the Culture of Disease' is a neat summary of anthrax and international trade; Bartrip, *The Home Office*, includes a chapter on anthrax which mainly focuses on Bradford as an example of regulation of occupational diseases; Holmes, *Spores, Plagues and History*, esp. pp. 90–1 and Swiderski, *Anthrax: A History*, esp. pp. 13–26, examine incidences of anthrax in Bradford within their biographies of the disease, but these summarize the existing secondary literature. Jones, *Death in a Small Package*, pp. 81–127, also examines Bradford within a biography of anthrax but contributes primary research. Jones and Teigen, 'Anthrax in Transit' analyses how knowledge of the transmission of anthrax travelled to three locations: Massachusetts, Glasgow and Bradford. The three PhD theses are Alvin, 'Medical Treatment and Care', p. 277; Stark, 'Industrial Illness', recently published as a book, J. Stark, *The Making of Modern Anthrax*,

1875–1920: Uniting Local, National and Global Histories of Disease (London: Pickering & Chatto, 2013); and Carter, 'Anthrax in Kidderminster'.

26. For example, N. F. Cantor, *In the Wake of the Plague: Black Death and the World it Made* (New York, Harper Collins, 2002).

27. K. C. Carter, 'The Koch–Pasteur Dispute on Establishing the Cause of Anthrax', *Bulletin of the History of Medicine*, 62 (1988), pp. 42–57, on pp. 46, 54.

28. B. Latour, *The Pasteurization of France*, trans. A. Sheridan and J. Law (Cambridge, MA, and London: Harvard University Press, 1988), pp. 87–90.

29. Holmes, *Spores, Plagues and History*; Swiderski, *Anthrax: A History*. See also Metcalfe, 'The History of Woolsorters' Disease', pp. 492–3; Nicoll and Maynard, 'One Hundred Years of Anthrax', p. 95.

30. J. Guillemin, *Anthrax: The Investigation of a Deadly Outbreak* (Berkeley, CA, and London: University of California Press, 1999).

31. 'Letter to the Editor', *Bradford Observer*, letter dated 7 March 1878, Bell Newspaper cuttings, file 1, p. 4.

32. T. Legge, 'The Milroy Lectures on Industrial Anthrax', *Lancet*, 1 (1905), pp. 690–1, cited in Mortimer and Melling, "The Contest between Commerce and Trade", p. 233.

33. Ibid., pp. 222, 232–3.

34. Carter, 'Anthrax in Kidderminster', pp. 194, 196.

35. K. Laybourn and D. James, 'Introduction', in K. Laybourn and D. James (eds), *The Rising Sun of Socialism: The Independent Labour Party in the Textile District of the West Riding of Yorkshire between 1890 and 1914* (West Yorkshire: West Yorkshire Archive Service, 1991), pp. x–xii, on p. x.

36. K. Ittman, *Work, Gender and Family in Victorian England* (New York: New York University Press, 1995), pp. 40–1; D. James, *Bradford* (Halifax: Ryburn Publishing, 1990), p. 77.

37. Bartrip, *The Home Office*, pp. 238–9.

38. James, *Bradford*, p. 52, cited in Stark, 'Industrial Illness', p. 16.

39. 'Strike of Woolsorters at Harden', *Bradford Observer*, 10 May 1880, Bell Newspaper cuttings, file 1, p. 25.

40. J. Spear, 'Appendix A. No. 8. On the so-called "Woolsorters' Disease" as observed at Bradford and in Neighbourhood Districts in the West Riding of Yorkshire', in *Supplement to the Tenth Annual Report of the Local Government Board, containing the Report of the Medical Officer for 1880* (London: HMSO, 1881), pp. 66–135.

41. Untitled, *Bradford Observer*, 8 May 1880, Bell Newspaper cuttings, file 1, p. 24.

42. 'Woolsorters' Disease', *Bradford Observer*, 6 June 1880, Bell Newspaper cuttings, file 1, p. 32.

43. *The Newspaper Press Directory and Advertisers' Guide, Containing Full Particulars of Every Newspaper, Magazine, Review, and Periodical Published in the United Kingdom and the British Isles with the Newspaper Map of the United Kingdom, the Principal Continental and American Papers, and a Directory of the Class Papers and Periodicals, Twenty-Fifth Annual Issue* (London: C. Mitchell and Co. Advertising Contractors, 1880), p. 49; D. James, 'William Byles and the *Bradford Observer*', in D. G. Wright and J. A. Jowitt (eds), *Victorian Bradford: Essays in Honour of Jack Reynolds* (Bradford: City of Bradford Metropolitan Council, 1982), pp. 115–36, on pp. 126–7; James, *Bradford*, pp. 40, 95; See also Jones, *Death in a Small Package*, p. 110. In 1893 the Conservative Party won all three Parliamentary seats in Bradford. See James, *Bradford*, p. 65.

44. James, *Bradford*, p. 76.

45. Stark, 'Bacteriology in the Service of Sanitation', p. 347.

46. G. Firth, *A History of Bradford* (Chichester: Phillimore, 1997), pp. 64–9; James, *Bradford*, p. 59.
47. Ibid., p. 44.
48. Ibid., pp. 44–5, 76.
49. Jones and Teigen, 'Anthrax in Transit', p. 466.
50. For example, Bligh, *Dr. Eurich*, p. 110; Carter, 'Anthrax in Kidderminster', p. 33, Metcalfe, 'The History of Woolsorters' Disease', p. 490.
51. 'Deaths from Blood Poisoning in Bradford', letter from Low Moor to *Bradford Observer*, 25 February 1878, Bell Newspaper cuttings, file 1, p. 1.
52. Bligh, *Dr. Eurich*, p. 110; Metcalfe, 'The History of Woolsorters' Disease', p. 490; Bell Newscuttings, file 1, p. 1.
53. Stark, 'Industrial Illness', p. 60.
54. Bradford was granted city status in 1897, hence references to a town before this date.
55. F. W. Eurich, 'The History of Anthrax in the Wool Industry of Bradford, and of its Control (being a Paper Read before the Bradford Medico-Chirurgical Society on the Occasion of the Centenary of the Bradford Royal Infirmary)', *Lancet*, 1 (1926), pp. 57–60, 107–9, on p. 57.
56. 'Blood Poisoning in Bradford', *Bradford Observer*, no date, file 1, Bell Newspaper cuttings, p. 1.
57. Bartrip, *The Home Office*, p. 236. Mortimer and Melling, "The Contest between Commerce and Trade", notes the low number of fatalities in Britain after 1905, and includes a table of cases and mortality rates nationally from 1899 to 1939, pp. 232–3, 225.
58. Local Government Board, *Tenth Annual Report of the Local Government Board, 1880–81. Supplement Containing Report and Papers Submitted by the Board's Medical Officer on the Use and Influence of Hospitals for Infectious Disease* (London: HMSO, 1882), p. 79.
59. 'Letters to the Editor', *Bradford Observer*, 7 March 1878; *Bradford Observer*, 11 March 1878; *Bradford Observer*, 16 March 1878, Bell Newspaper cuttings, file 1, p. 6.
60. 'Letters to the Editor', *Bradford Observer*, 8 March 1878, Bell Newspaper cuttings, file 1, p. 5.
61. Untitled, *Bradford Observer*, 7 March 1878, Bell Newspaper cuttings, file 1, p. 5.
62. Untitled, *Bradford Observer*, 7 May 1880, Bell Newspaper cuttings, file 1, p. 22; Bartrip, *The Home Office*, p. 236, n. 16.
63. 'The Woolsorters' Disease', *Bradford Observer*, 5 March 1878, Bell Newspaper cuttings, file 1, p. 2.
64. Untitled, unlabelled paper, 5 March 1878, Bell Newspaper cuttings, file 1, p. 3. It seems very likely that this was an article in the *Bradford Observer* as there was what looks like a direct response to the article printed in the *Bradford Observer*, 'Letter to the Editor', letter dated 6 March 1878, Bell Newspaper cuttings, file 1, p. 4.
65. Bartrip, *The Home Office*, p. 239; 'The Woolsorters' Disease', *Bradford Observer*, 16 March 1878, Bell Newspaper cuttings, file 1, p. 8.
66. Ibid.
67. Bligh, *Dr Eurich*, pp. 109–10.
68. Ibid., pp. 115–6; Jones and Teigen, 'Anthrax in Transit', p. 466.
69. Mortimer and Melling, "The Contest between Commerce and Trade", p. 223.
70. Alvin, 'Medical Treatment and Care', p. 286.
71. Bradford Medico-Chirurgical Society, *Report of the Commission on Woolsorter's Disease* (Bradford, 1882), pp. 1, 2, 7, 9, 10.

72. E. T. Tibbits, *Medical Fashions in the Nineteenth Century, including a Sketch of Bacteriomania and the Battle of the Bacilli* (London: H. K. Lewis, 1884), reverse of title page. Tibbits is briefly mentioned in Bartrip, *The Home Office*, p. 242.

73. Letter from E. T. Tibbits, *The Times*, 27 August 1881, p. 4. See Jones, *Death in a Small Package*, pp. 74–5; Jones and Teigen, 'Anthrax in Transit', p. 478; and Stark, 'Industrial Illness', pp. 29–30, for other examples of Tibbits's arguments in the medical press.

74. Untitled, *Bradford Observer*, 4 December 1879, Bell Newspaper cuttings, file 1, p. 14.

75. 'Letter to the Editor', *Bradford Observer*, letter dated 8 December 1879, Bell Newspaper cuttings, file 1, p. 16.

76. 'More Fatalities from Woolsorters' Disease', *Bradford Daily Telegraph*, 7 May 1880, Bell Newspaper cuttings, file 1, p. 23.

77. Stark, 'Industrial Illness', p. 56.

78. 'More Fatalities from Woolsorters' Disease', *Bradford Daily Telegraph*, 7 May 1880, Bell Newspaper cuttings, file 1, p. 23.

79. Bradford Medico-Chirurgical Society, *Report*, p. 11.

80. Ibid.

81. Bradford Medico-Chirurgical Society Minute Book, 1874–84, 4 April 1882, WYASB, 40D89/2.

82. Borough of Bradford, *Report on the Health of Bradford by the Medical Officer of Health, 1884* (Bradford: Bradford Borough Council, 1885), p. 60.

83. Tibbits, *Medical Fashions*, p. 38. Tomes cites an American who used the phrase 'bacteriomania' in 1885, a physician who was also fighting a 'rearguard action'. See Tomes, 'American Attitudes', p. 42, citing A. Jacobi, 'Inaugural Address', *Med. Rec.* (N.Y.), 27 (1885), pp. 169–74, on p. 172 for 'bacteriomania'.

84. Tibbits, *Medical Fashions*, p. 38; Tibbits challenged Pasteur, Klebs, Greenfield and Spear in the *Bradford Observer*, 29 August 1881, cited in Stark, 'Industrial Illness', p. 163. Tibbets challenged Greenfield in the *British Medical Journal* (19 February 1881), p. 293, cited in Jones and Teigen, 'Anthrax in Transit', pp. 478–9.

85. Tibbits, *Medical Fashions*, pp. 52–3.

86. Ibid., p. 39

87. Ibid., pp. 44–53.

88. Ibid., p. 55.

89. Ibid., p. 62.

90. Worboys, *Spreading Germs*, pp. 63, 137.

91. L. G. Stevenson, 'Science Down the Drain: On the Hostility of Certain Sanitarians to Animal Experimentation, Bacteriology and Immunology', *Bulletin of the History of Medicine*, 29 (1955), pp. 1–26, on p. 17.

92. Worboys, *Spreading Germs*, pp. 63, 137, 146.

93. Ibid., p. 206.

94. Ibid., pp. 205–6.

95. Ibid., p. 209.

96. Ibid., pp 209–10, 228, 231.

97. Jones, *Death in a Small Package*, pp. 103–4.

98. Bligh, *Dr. Eurich*, p. 117.

99. 'Letters to the Editor', *Bradford Observer*, 8 December 1879, Bell Newspaper cuttings, file 1, p. 16.

100. Ibid.

101. 'The Death from the Woolsorters' Disease. To the Editor', *Bradford Observer*, 7 May 1880, Bell Newspaper cuttings, file 1, p. 21.

102. Untitled, *Bradford Observer*, 12 July 1881, Bell Newspaper cuttings, file 1, p. 61.

103. J. Spear, 'The "Woolsorters' Disease", or Anthrax Fever, read May 4th 1881', *Transactions of the Epidemiological Society of London*, 4 (1875–81), pp. 610–30, on p. 613; Spear, 'On the so-called "Woolsorters' Disease"', p. 78.

104. Ibid., p. 103; Spear, 'The "Woolsorters' Disease"', p. 613.

105. Jones and Teigen, 'Anthrax in Transit', p. 480; C. Steedman, *Dust* (Manchester: Manchester University Press, 2000), p. 23.

106. Alvin, 'Medical Treatment and Care', p. 277.

107. Eurich, 'The History of Anthrax', p. 57.

108. 'Letter to the Editor', *Bradford Observer*, letter dated December 1879, Bell Newspaper cuttings, file 1, p. 18.

109. Untitled, *Yorkshireman*, 20 December 1879, Bell Newspaper cuttings, file 1, p. 18.

110. Ibid.

111. 'Letter to the Editor', *Bradford Daily Telegraph*, letter dated 10 May 1880, Bell Newspaper cuttings, file 1, p. 25. Eurich, 'The History of Anthrax', p. 58.

112. 'The Recent Death from Woolsorters' Disease', *Bradford Daily Telegraph*, 10 June 1880, Bell Newspaper cuttings, file 1, p. 32.

113. N. Tomes, 'The Private Side of Public Health: Sanitary Science, Domestic Hygiene, and the Germ Theory, 1870–1900', *Bulletin of the History of Medicine*, 64 (1990), pp. 509–39.

114. 'Letter to the Editor', *Bradford Daily Telegraph*, letter dated 10 May 1880, Bell Newspaper cuttings, file 1, p. 25.

115. 'A Word on "Woolsorters' Diseases" by a Sorter', *Yorkshireman*, 3 July 1880, p. 10.

116. 'Letter to the Editor', *Bradford Observer*, letter dated 3 October 1881, Bell Newspaper cuttings, file 1, p. 64.

117. 'The Death from Woolsorters' Disease in Bradford. Inquest', *Bradford Daily Telegraph*, 3 July 1888, Bell Newspaper cuttings, file 1, p. 95.

118. Borough of Bradford, *Report on the Health of Bradford, for the year 1880*, p. 31. According to an article in the *Lancet* the intestinal variety was only acquired from eating raw or partially cooked meat: 'Anthrax at Bradford', *Lancet*, 2 (1908), pp. 1453–4, on p. 1453. However, presumably spores could land on prepared food.

119. This trend was also noted by Carter, 'Anthrax in Kidderminster', p. 197, in his discussion of the Kidderminster press.

120. James, *Bradford*, p. 110.

121. Jones, *Death in a Small Package*, p. 86.

122. Spear, 'On the so-called "Woolsorters' Disease"', pp. 78–9.

123. Spear, 'The "Woolsorters' Disease"', p. 613.

124. National Union of Woolsorters' Minutes, 1898–1971, 21 June 1901, WYASB, 21/D82/1/1.

125. Ittman, *Work, Gender and Family*, p. 41.

126. 'Woolsorters' Disease', *Bradford Observer*, 1 July 1884, Bell Newspaper cuttings, file 1, p. 66.

127. Untitled, *Bradford Observer*, 7 June 1888, Bell Newspaper cuttings, file 1, p. 93; undated, untitled, unlabelled newspaper article, Bell Newspaper cuttings, file 1, p. 93, but refers to the same inquest as occurring 'yesterday'.

128. T. Carter, 'The Dissemination of Anthrax from Imported Wool: Kidderminster 1900–14', *Occupational Health and Environmental Medicine*, 61 (2004), pp. 103–7, on p. 104.

129. 'Letter from "A Woolsorter"', *Bradford Observer*, 11 May 1880, Bell Newspaper cuttings, file 1, p. 25.

130. Ibid.

131. 'Deaths from Blood Poisoning in Bradford. To the Editor', *Bradford Observer*, 1 March 1878, Bell Newspaper cuttings, file 1, p. 1.

132. 'To the Editor', *Bradford Observer*, letter dated 2 March 1878, Bell Newspaper cuttings, file 1, p. 1.

133. 'The Woolsorters' Disease. Inquest at Low Moor', *Bradford Observer*, 5 March 1878, Bell Newspaper cuttings, file 1, p. 2.

134. Untitled, *Bradford Daily Telegraph*, 5 March 1878, Bell Newspaper cuttings, file 1, p. 3.

135. 'Letters to the Editor', *Bradford Observer*, 6 March 1878, Bell Newspaper cuttings, file 1, p. 4.

136. 'The Woolsorters' Disease. Letters to the Editor', 7 March 1878, Bell Newspaper cuttings, file 1, p. 4.

137. 'The Woolsorters' Disease', *Bradford Observer*, 16 March 1878, Bell Newspaper cuttings, file 1, p. 8.

138. 'Letters to the Editor', *Bradford Observer*, 13 December 1879, Bell Newspaper cuttings, file 1, p. 18; *Bradford Observer*, 11 May 1880, Bell Newspaper cuttings, file 1, p. 25; *Bradford Observer*, May 1880, Bell Newspaper cuttings, file 1, p. 27. There is also a letter from an employer in Constantinople regarding the dangers of the wool, and the dung attached to the fleeces, *Bradford Observer*, 5 January 1880, Bell Newspaper cuttings, file 1, p. 20.

139. J. H. Bell, 'Notebook Belonging to Dr. Bell, Written during the Great Anthrax Epidemic of 50 Years Ago', WYASB, deed box 15, case no. 5.

140. T. Legge, 'The Milroy Lectures on Industrial Anthrax', *British Medical Journal* (1905), 1, pp. 529–31, 589–93, 641–3, cited in Metcalfe, 'The History of Woolsorters' Disease', p. 491.

141. Bartrip, *The Home Office*, p. 240.

142. Jones and Teigen, 'Anthrax in Transit', p. 481.

143. Stark, 'Industrial Illness', p. 43.

144. Melling, 'Beyond a Shadow of a Doubt', p. 432; Steedman, *Dust*, pp. 20–1.

145. Stark, 'Bacteriology in the Service of Sanitation', pp. 348–51.

146. Bell, 'Notebook'.

147. 'Letter to the Editor', *Bradford Observer*, undated, Bell Newspaper cuttings, file 1, p. 18.

148. 'Letter to the Editor', *Bradford Observer*, undated letter, though with one dated 2 March 1878, Bell Newspaper cuttings, file 1, p. 1.

149. 'Letter to the Editor', *Bradford Observer*, letter dated 11 May 1881, Bell Newspaper cuttings, file 1, p. 25.

150. Jones and Teigen, 'Anthrax in Transit', p. 466.

151. Bradford Medico-Chirurgical Society Minute Book, 1874–84, 19 September 1883, WYASB, 40D89/2.

152. Untitled, unlabelled newspaper, 3 March 1878, Bell Newspaper cuttings, file 1, p. 3.

153. Untitled, *Bradford Daily Telegraph*, 5 March 1878, Bell Newspaper cuttings, file 1, p. 3.

154. 'Letter to the Editor', *Bradford Observer*, letter dated 8 March 1878, Bell Newspaper cuttings, file 1, p. 5; 'The Woolsorters' Disease', *Bradford Observer*, 16 March 1878, Bell Newspaper cuttings, file 1, p. 8.

155. Ibid.

156. 'The Woolsorters' Plague', *Yorkshireman*, December 1879, Bell Newspaper cuttings, file 1, p. 15; Untitled, *Bradford Observer*, 8 May 1880, Bell Newspaper cuttings, file 1, p. 24.

157. Untitled, *Bradford Observer*, 7 May 1880, Bell Newspaper cuttings, file 1, p. 22.

158. 'The Woolsorters' Disease. To the Editor', *Bradford Observer*, 3 October 1881, file 1, Bell Newspaper cuttings, p. 64.

159. A safe vaccine for humans was not administered to wool workers until the 1950s, although the serum mentioned above was available. See Wilkinson, 'Anthrax', p. 583; Alvin, 'Medical Treatment and Care', p. 286.

160. 'The Woolsorters' Disease. To the Editor', *Bradford Observer*, 3 October 1881, file 1, Bell Newspaper cuttings, p. 64.

161. Carter, 'The Koch–Pasteur Dispute', pp. 56–7.

162. *The Times*, 3 June 1881, p. 5.

163. N. Pemberton and M. Worboys, *Rabies in Britain: Dogs, Disease and Culture, 1830–2000* (Basingstoke: Palgrave Macmillan, 2012), pp. 4, 102, 111–12. Another dog-bite victim who came forward for treatment later than the others was treated in London as anti-vivisectionists ordered his treatment, p. 115.

164. *The Times*, 13 March 1886, p. 7.

165. Pemberton and Worboys, *Rabies in Britain*, pp. 112, 115.

166. Ibid., p. 108.

167. Ibid., p. 115.

168. K. Figlio, 'What is an Accident?', in P. Weindling (ed.), *The Social History of Occupational Health* (London: Croom Helm, 1985), pp. 180–206, on p. 198.

169. Stark, 'Industrial Illness', p. 149.

170. Carter, 'Anthrax in Kidderminster', pp. ii, 194.

171. 'Anthrax and the Compensation Act', *Yorkshire Daily Observer*, 16 May 1905, Bell Newspaper cuttings, file 2, p. 74; 'Death from Anthrax. Bradford Widow Claims Compensation', unlabelled newspaper, undated although filed with cuttings from April 1905, Bell Newspaper cuttings, file 2, p. 72; 'Saltaire Anthrax Case. Widow Gets £242 Compensation', *Bradford Daily Telegraph*, 28 April 1905, Bell Newspaper cuttings, file 2, p. 72; 'The Anthrax Terror. Bradford Claim for Compensation', *Yorkshire Daily Observer*, 15 May 1905, Bell Newspaper cuttings, file 2, p. 73; 'Workman's Death from Anthrax. A Claim for Compensation', *Bradford Daily Telegraph*, 7 June 1905, Bell Newspaper cuttings, file 2, p. 74. Compensation was partially rescinded in one case as a widow lived with another man, see Stark, 'Industrial Illness', p. 154.

172. Lord Macnaghten's assertion in *Fenton v. Thorley & Co. Ltd* [1903], cited in P. Bartrip, *Workmen's Compensation in Twentieth Century Britain: Law, History and Social Policy* (Aldershot: Avebury, 1987), pp. 26, 53.

173. J. Moses, 'Contesting Risk: Specialist Knowledge and Workplace Accidents in Britain, Germany, and Italy, 1870–1920', in K. Brückweh, D. Schumann, R. F. Wetzell and B. Ziemann, *Engineering Society: The Role of the Human and Social Sciences in Modern Societies, 1880–1980* (Basingstoke: Palgrave, 2012), pp. 59–78, on pp. 67–8.

174. Bartrip, *Workmen's Compensation*, pp. ix, 2–3, 8.

175. 'Another Death from Woolsorters' Disease in Bradford', *Bradford Daily Telegraph*, 7 May 1880, Bell Newspaper cuttings, file 1, p. 21.

176. 'What Will People Want Compensation for Next?', *Yorkshireman*, 11 September 1880, p. 164.

177. 'The Pillory [a column]. Woolsorters' Disease', *Yorkshireman*, 12 June 1880, p. 373.

178. 'A Word on "Woolsorters' Diseases" by a Sorter', *Yorkshireman*, 3 July 1880, p. 10.

179. 'The Recent Death from Woolsorters' Disease', *Bradford Daily Telegraph*, 10 June 1880, Bell Newspaper cuttings, file 1, p. 32.

180. Stark, 'Industrial Illness', p. 142.

181. Bartrip, *Workmen's Compensation*, p. 12.

182. Untitled, unlabelled, undated newspaper, filed with 1899, Bell Newspaper cuttings, file 2, p. 36.

183. 'Is Anthrax an "Accident"', *Bradford Observer*, 25 July 1900, Bell Newspaper cuttings, file 2, p. 38; see also Bartrip, *Workmen's Compensation*, p. 10.

184. 'Is Anthrax an "Accident"', unlabelled, undated newspaper, filed with July 1900, Bell Newspaper cuttings, file 2, p. 39A.

185. Tomes, *Gospel of Germs*.

186. Worboys, *Spreading Germs*, pp. 160–1.

187. Ibid., pp. 135–7.

188. Stark, 'Bacteriology in the Service of Sanitation', p. 355.

189. Stark, 'Industrial Illness', p. 81; *Annual Reports of the Chief Inspector of Factories* 1900–39, cited in Mortimer and Melling, "The Contest between Commerce and Trade", p. 225.

190. See especially Mortimer and Melling, "The Contest between Commerce and Trade"; Stark, 'Industrial Illness', pp. 156–60.

191. Jones, *Death in a Small Package*, p. 113; Stark, 'Industrial Illness', p. 157–9.

192. Mortimer and Melling, "The Contest between Commerce and Trade", pp. 228, 232; Jones, *Death in a Small Package*, p. 113.

4 Anthrax in London: Leather, Zoo Keeping and Shaving Brushes, 1882–1932

1. Quote from a 'popular penny-weekly', in 'Tanning as a "Dangerous" Trade', *Leather World*, 3 October 1912, pp. 1–2.

2. J. Spear, 'On the Occurrence of Anthrax amongst Persons Engaged in the London Hide and Skin Trades', in *Twelfth Annual Report of the Local Government Board: Supplement containing the Report of the Medical Officer for 1882* (London: HMSO, 1883), pp. 98–131, on pp. 105–7.

3. C. Ponder, *A Report to the Worshipful Company of Leathersellers on the Incidence of Anthrax amongst those Engaged in the Hide, Skin and Leather Industries, with an Inquiry into Certain Measures Aiming at its Prevention* (London: Worshipful Company of Leathersellers, 1911), p. 24.

4. 'Industrial Anthrax', *British Medical Journal* (25 March 1905), p. 668.

5. Statistics derived from 'Table 1: Incidents of Anthrax and Fatality Rates 1899–1939', compiled from *Annual Reports of the Chief Inspector of Factories* 1900–39, in Mortimer and Melling, "The Contest between Commerce and Trade", p. 225.

6. Ponder, *Report to the Worshipful Company of Leathersellers*, p. 20.

7. Statistics derived from 'Table 1: Incidents of Anthrax and Fatality Rates 1899–1939', compiled from *Annual Reports of the Chief Inspector of Factories* 1900–39, in Mortimer and Melling, "The Contest between Commerce and Trade", p. 225.

8. Ibid., p. 225.

9. W. H. Hamer, *Report of the Medical Officer on Anthrax in London* (London: London County Council, 1894), pp. 1, 12, 14; Ponder, *Report to the Worshipful Company of Leathersellers*, p. 2.

10. N. Davies-Colley, 'Report of Cases of Anthrax or Malignant Pustule under the Care of Mr. Davies-Colley, with Notes upon the Treatment of this Affection by the External and Internal Use of Ipecacuanha', *Guy's Hospital Reports*, 47 (1890), pp. 1–20.

11. Stark, 'Industrial Illness', p. 77.

12. 'A Victim of Anthrax', *South London Press*, 30 March 1907, p. 6.

13. Stark, 'Industrial Illness', p. 10.

14. Jones, *Death in a Small Package*, briefly mentions the risk among hide tanners in France and Germany in the nineteenth century, pp. 50, 114, and in the US and France in the early twentieth century, pp. 114, 116; Holmes, *Spores, Plagues and History*, pp. 1, 9–10, 23, 88; Swiderski, *Anthrax: A History*, pp. 1, 32.

15. Bartrip, *The Home Office*, pp. 243–4; Stark, 'Industrial Illness', p. 113–4.

16. Bartrip, *The Home Office*, pp. 244–5.

17. Jones, *Death in a Small Package*; Holmes, *Spores, Plagues and History*; Swiderski, *Anthrax: A History*.

18. International Anthrax Commission, *Memorandum Circulated by the British Representative* (London: HMSO, 1922), International Labour Organization Archive (hereafter ILOA), Hy 501/1/2.

19. Mortimer and Melling, "The Contest between Commerce and Trade", p. 231; Carter and Melling, 'Trade, Spores, and the Culture of Disease'.

20. Mortimer and Melling, "The Contest between Commerce and Trade", p. 231.

21. Bartrip, *The Home Office*, pp. 256–7.

22. 'Anthrax in Bermondsey', *British Medical Journal* (17 March 1883), pp. 527–8.

23. Spear, 'On the Occurrence of Anthrax', p. 98.

24. Ibid., pp. 98–100.

25. Ponder, *Report to the Worshipful Company of Leathersellers*, pp. 12–13, 30–3, 50.

26. Ibid., p. 14.

27. Ibid., pp. 15–16.

28. Ibid., pp. 16–18, 26–7.

29. 'Anthrax', *British Medical Journal* (14 June 1884), p. 1161.

30. Hamer, *Anthrax in London*, pp. 8–25.

31. Davies-Colley, 'Report of Cases of Anthrax', pp. 1–20.

32. Spear, 'On the Occurrence of Anthrax', p. 131.

33. Stark, 'Bacteriology in the Service of Sanitation', p. 351.

34. Davies-Colley, 'Report of Cases of Anthrax', pp. 1–16.

35. 'Anthrax at Guy's Hospital', *British Medical Journal* (11 November 1893), p. 1068.

36. Hamer, *Anthrax in London*, p. 1.

37. 'Anthrax in London', *British Medical Journal* (18 August 1894), p. 380.

38. 'Death from Anthrax', *British Medical Journal* (5 October 1895), p. 852; 'Anthrax in Bermondsey', *British Medical Journal* (8 February 1896), p. 356.

39. 'The Port of London', 26 November 1901, p. 13; 'The Public Health', 10 July 1902, p. 2; 'The Public Health', 12 March 1903, p. 3; 'The Port of London', 26 July 1906, p. 12; 'Inquests', 23 March 1907, p. 8; 'Death from Anthrax', 1 July 1909, p. 16; 'Death from Anthrax', 24 August 1909, p. 10, all from *The Times*.

40. 'Anthrax in London', *The Times*, 24 December 1913, p. 8.

41. 'Anthrax Infection', *The Times*, 7 January 1914, p. 11. Another death at a wharf was recorded in 1917, but the inquest decided that this was an accidental death but with no neglect on the part of the firm involved.

42. 'Death from Anthrax', *The Times*, 26 May 1920, p. 18.
43. 'Three Fatal Anthrax Cases', *Leather World*, 17 April 1913, p. 302.
44. 'China Hide Trade', *Leather World*, 18 July 1912, pp. 1–2; 'The Deadly Anthrax', 10 April 1913, pp. 269–70; 'Another Anthrax Fatality', *Leather World*, 17 April 1913, p. 308.
45. 'Labour Unrest in the Leather Trade', *Leather World*, 22 May 1913, pp. 397–8.
46. 'Tanning as a "Dangerous" Trade', pp. 1–2.
47. 'The Deadly Anthrax', *Leather World*, 10 April 1913, pp. 269–70.
48. 'China Hide Trade', *Leather World*, 18 July 1912, pp. 1–2; 'The Deadly Anthrax', 10 April 1913, pp. 269–70; 'Anthrax and Hide Sterilisation', *Leather World*, 24 April 1913, p. 323; 'A Study of Anthrax Germs', 14 December 1911, pp. 956–7; 'The Disinfection of Hides', *Leather*, 13 April 1911, p. 275.
49. 'A Study of Anthrax Germs', *Leather*, 7 December 1911, pp. 644–5 and 14 December 1911, pp. 956–7.
50. 'The Disinfection of Hides', *Leather*, 6 April 1911, p. 256 and 13 April 1911, p. 275.
51. 'China Hide Trade', *Leather World*, 18 July 1912, pp. 1–2; 'Tannery Effluent and Anthrax', *Leather World*, 15 August 1912, p. 616.
52. For example, 'Dr Eurich on Anthrax Prevention', *Leather World*, 22 August 1912, p. 633; 'Fatal Anthrax Case', *Leather World*, 5 September 1912 p. 666; 'Deaths from Anthrax', *Leather World*, 12 December 1912, p. 941; 'More Anthrax Cases', *Leather World*, 19 December 1912, p. 955; 'Bradford Inquest', *Leather World*, 7 March 1912, p. 178.
53. For example 'Anthrax', *Leather World*, 6 June 1912, p. 422; 'Deaths from Anthrax', *Leather World*, 12 December 1912, p. 941.
54. For example, 'Anthrax in Bermondsey', *Leather World*, 8 August 1912, p. 587.
55. 'Deaths from Anthrax', *Leather World*, 12 December 1912, pp. 941–2.
56. 'The Deadly Anthrax', *Leather World*, 10 April 1913, pp. 269–70 (this is the lead article on the front page).
57. Ibid., pp. 269–70.
58. Ibid.
59. Ponder, *Report to the Worshipful Company of Leathersellers*, pp. 37–9.
60. 'The Deadly Anthrax', *Leather World*, 10 April 1913, pp. 269–70.
61. 'The Disinfection of Hides', *Leather World*, 13 April 1911, p. 275.
62. Ponder, *Report to the Worshipful Company of Leathersellers*, p. 20.
63. Ibid., pp. 24–6.
64. R. Gorski, 'Health and Safety aboard British Merchant Ships: The Case of First Aid Instruction, 1881–1908', in R. Gorski (ed.), *Maritime Labour: Contributions to the History of Work at Sea, 1500–2000* (Amsterdam: Aksant, 2007), pp. 119–40.
65. Ponder, *Report to the Worshipful Company of Leathersellers*, p. 26.
66. Ibid., p. 40.
67. 'Death from Anthrax', *Leather World*, 2 May 1912, p. 322.
68. Ponder, *Report to the Worshipful Company of Leathersellers*, pp. 26–9.
69. 'Death from Anthrax', *Leather World*, 2 May 1912, p. 322.
70. Ponder, *Report to the Worshipful Company of Leathersellers*, pp. 41–6, 73–88.
71. J. B. Andrews, *Anthrax as an Occupational Disease: Bulletin of the United States Bureau of Labor Statistics, Industrial Hygiene Series*, 10 (1917), p. 118.
72. Ponder, *Report to the Worshipful Company of Leathersellers*, pp. 55–63.
73. Ibid., pp. 10, 37–9.
74. Ibid., pp. 41–6.
75. Ibid., p. 40.

76. Ibid.
77. T. Legge, 'The Milroy Lectures on Industrial Anthrax', *British Medical Journal* (25 March 1905), pp. 641–3.
78. Ponder, *Report to the Worshipful Company of Leathersellers*, p. 40.
79. Ibid., p. 26.
80. W. A. Willis, *The Workmen's Compensation Act, 1906* (London: Butterworth and Co. and Shaw and Sons, 1913), pp. 134–5.
81. D. Wilson, *Dockers: The Impact of Industrial Change* (London: Fontana, 1972), pp. 17, 22. Wilson cites P. Quennell (ed.), *Mayhew's London* (1969), p. 568.
82. Wilson, *Dockers*, p. 17, 22–3.
83. Ibid., pp. 18, 21.
84. Ibid., pp. 18–23, 82.
85. B. Webb, 'Dock Life in the East End of London', *Nineteenth Century* (September 1887), cited in Wilson, *Dockers*, pp. 21–2.
86. Ibid., p. 21.
87. Wilson, *Dockers*, pp. 42–3.
88. E. L. Taplin, *Liverpool Dockers and Seamen, 1870–1890* (Hull: University of Hull Publications, 1974), pp. 6–9.
89. Ibid.
90. Wilson, *Dockers*, pp. 58, 98.
91. Ibid., pp. 63–4, 73.
92. Mr Astor to Mr McKenna, Written Answers, Hansard, House of Commons (HC) 20 March 1911.
93. See Mortimer and Melling, "The Contest between Commerce and Trade", p. 225 for the statistics of the number of cases derived from the *Annual Reports of the Chief Inspector of Factories* 1900–39.
94. 'Skin Diseases in Relation to Industry', *British Medical Journal* (13 August 1932), pp. 292–4.
95. Bartrip, *The Home Office*, p. 248.
96. 'The World's Labour Charter', *The Times*, 27 October 1920, p. 11.
97. 'Next World Labour Conference', *The Times*, 20 January 1921, p. 10.
98. International Labour Office, Advisory Committee on Anthrax, First Sitting, 5 December 1922, ILOA, Hy 550/2/1.
99. International Anthrax Commission, *Memorandum Circulated By the British Representative*, ILOA, Hy 501/1/2, p. 7.
100. International Labour Office, Advisory Committee on Anthrax, First Sitting, 5 December 1922, ILOA, Hy 550/2/1.
101. International Labour Office, Advisory Committee on Anthrax, December 1922, Correction of Minutes, Eighth Meeting, ILOA, Hy 550/1/11.
102. Secretary, Anthrax Commission, to L. Leckie, Amalgamated Union of Upholsterers, 19 December 1932, ILOA, Hy 550/3/25/2.
103. International Anthrax Commission, *Memorandum Circulated By the British Representative*, ILOA, Hy 501/1/2, p. 8.
104. Letter to E. Collis, 8 September 1924, ILOA, Hy 550/6.
105. M. Delevigne to H. B. Butler, 22 May 1924, ILOA, Hy 550/6.
106. Butler to Delevigne, 28 May 1924, ILOA, Hy 550/6.
107. Butler to Carozzi, 19 July 1923, ILOA, Hy 550/6.
108. Mortimer and Melling, "The Contest between Commerce and Trade", pp. 227, 230.

109. 'Cases of Anthrax in London', *The Times*, 8 December 1922, p. 9.

110. Mortimer and Melling, "The Contest between Commerce and Trade", p. 230.

111. Delevigne to Butler, 10 May 1924, ILOA, Hy 550/6.

112. As discussed on pp. 97–8.

113. International Labour Office, Minutes of the 23rd Session of the Governing Body, June 1924, p. 275.

114. T. Legge to A. Thomas, 30 April 1924; Butler to Delevigne, 15 May 1924, ILOA, Hy 550/6; J. Melling and C. Sellers, 'Objective Collectives? Transnationalism and "Invisible Colleges" in Occupational and Environmental Health from Collis to Selikoff', in Sellers and Melling, *Dangerous Trade*, pp. 113–25, on p. 114.

115. Ibid., p. 117.

116. F. Coutts, *Report to the Local Government Board on an Inquiry into Cases of Anthrax (Malignant Pustule or External Anthrax) Suspected to be due to Infected Shaving Brushes* (London: Local Government Board, 1917).

117. Anthrax: Enquiry by Professor Ottolenghi, 1928–29, League of Nations Archives, Industrial Hygiene, R5880, 8A/3856/3856.

118. Anthrax: Correspondence with M. Andrew Poulson, 1928, League of Nations Archives, Industrial Hygiene, R5880, 8A/8921/3856/1.

119. International Labour Office, 'The Prevention of Anthrax: Draft Regulations for Protection Against Infection by Anthrax in the Hides and Skins Industry', in International Labour Office, *The International Labour Code, 1939* (Montreal: International Labour Office, 1941), pp. 590–9, on pp. 596–9.

120. Carter and Melling, 'Trade, Spores, and the Culture of Disease', pp. 68–9.

121. Mr Citrine, General Secretary, Trades Union Congress General Council, to Albert Thomas, ILO, 10 December 1929; Draft Letter to Mr Citrine, ILOA, Hy 550/3/26/6.

122. Letter to Mr Citrine, 17 January 1930, ILOA, Hy 550/3/25/6.

123. Letter from Mixed Committee to Dr Cummings, Surgeon General, Director Public Health Service, US, 4 June 1932, ILOA, Hy 550/4.

124. Letter from E. Collis to L. Carozzi, 29 May 1929, ILOA, Hy 550/2/19.

125. Mortimer and Melling, "The Contest between Commerce and Trade", p. 232.

126. 'Anthrax in London Horses', *British Medical Journal* (8 June 1895), p. 1285; 'Anthrax in England', *British Medical Journal* (26 March 1887), p. 687.

127. 'Anthrax at the Parcels Post Office', *British Medical Journal* (21 May 1898), p. 1346.

128. 'A Laboratory Attendant's Death', *The Times*, 10 July 1911.

129. D. Hunter, *The Diseases of Occupations* (London: English Universities Press, 1955), pp. 199, 620–38.

130. 'Anthrax at the Zoo', *The Times*, 13 January 1927, p. 9.

131. H. H. Scott, 'Report on the Deaths Occurring in the Society's Gardens during the Year 1926', *Proceedings of the Zoological Society of London*, 97 (1927), pp. 173–98, on pp. 179–84.

132. *ODNB* [accessed 4 February 2013].

133. H. H. Scott, *Tuberculosis in Man and Lower Animals: A Study in Comparative Pathology*, Medical Research Council, Special Report Series, No. 149 (1930); H. H. Scott, 'Tuberculosis in Captive Wild Animals as Compared and Contrasted with the Disease in Man', *Proceedings of the Royal Society of Medicine*, 20 (1927), pp. 197–204.

134. Scott, 'Report on the Deaths'.

135. Zoological Society Minutes of Council, 1926–9, 19 January 1927, Zoological Society of London Archives, GB 0814 FAAA.

136. 'Goat's Hair Brushes', *The Times*, 27 June 1917, p. 3; Coutts, *Report to the Local Government Board*, pp. 1–5.
137. Coutts, *Inquiry into Cases of Anthrax*, pp. 3–14.
138. Ibid., p. 10.
139. 'Shaving brushes infected with Anthrax Spores', *British Medical Journal* (7 March 1925), p. 470; Coutts, *Inquiry Into Cases of Anthrax*, pp. 14–15, 21.
140. Ibid., pp. 6–14.
141. Ibid., p. 13.
142. Ibid., p. 18.
143. 'Anthrax in Shaving Brushes', *The Times*, 25 August 1916, p. 3.
144. 'Goat's Hair Brushes', *The Times*, 6 June 1917, p. 3.
145. 'Anthrax in Shaving Brushes', *The Times*, 6 January 1920, p. 12.
146. 'Healthy London', *The Times*, 24 July 1917, p. 3.
147. 'Anthrax from Shaving-Brushes', *British Medical Journal* (5 February 1921), p. 207; 'Anthrax from Shaving Brushes', *British Medical Journal* (22 October 1921), p. 662.
148. 'Goat's Hair Brushes', *The Times*, 27 June 1917, p. 3.
149. 'Shaving Brushes: New Official Warning', *The Times*, 31 January 1921, p. 7.
150. 'Anthrax-Infected Shaving Brushes', *British Medical Journal* (14 March 1925), p. 532.
151. Mortimer and Melling, "The Contest between Commerce and Trade", p. 224.
152. Jones, *Death in a Small Package*, pp. 121–4.
153. BBC News, 'Anthrax Outbreak Hits Bangladesh Leather and Meat Sectors', 13 October 2010, at www.bbc.co.uk/news/business-11451570 [accessed 30 October 2012].
154. World Organization for Animal Health, World Health Organization and Food and Agriculture Organization of the United Nations, *Anthrax in Humans and Animals* (Geneva: World Health Organization, 2008), esp. p. 28.
155. Jones, *Death in a Small Package*, p. xvi.
156. Steedman, *Dust*, pp. 25–8. Collectors of shaving brushes may also have similar fears. A contributor to the shaving forum, 'Straight Razor Place', who is an operating room technician, started a thread of online conversation which discussed early twentieth-century brushes which were imprinted with the word 'sterilized' and the dangers of anthrax for current collectors. See http://straightrazorplace.com/brushes/58660-anthrax-shaving-brushes-1925-article.html [accessed 30 October 2012].

5 Charity, Compensation and Carriers: The Development of Blame and Responsibility in Response to Epidemics

1. E. Tracey Archer, Correspondence to the Editor of the *Malton Messenger*, 19 November 1932, File on Water, North Yorkshire Record Office, Typhoid Relief Fund, ZPB (hereafter NYRO TRF).
2. See for example, J. Robins, *The Miasma: Epidemic and Panic in Nineteenth Century Ireland* (Dublin: Institute of Public Administration, 1995); F. M. Snowden, *Naples in the Time of Cholera, 1884–1911* (Cambridge: Cambridge University Press, 1995); R. E. McGrew, *Russia and the Cholera 1823–1832* (Madison and Milwaukee, WI: The University of Wisconsin Press, 1965); R. J. Evans, *Death in Hamburg: Society and Politics in the Cholera Years, 1830–1910* (Oxford: Clarendon Press, 1987); F. Delaporte, *Disease and Civilisation: The Cholera in Paris, 1832*, trans. A. Goldhammer (Cambridge: MA, and London, The MIT Press, 1986); C. E. Rosenberg, *The Cholera Years: The United*

States in 1832, 1849, and 1866 (Chicago, IL, and London: University of Chicago Press, 1962); H. Prescott, 'Sending their Sons into Danger: Cornell University and the Ithaca Typhoid Epidemic of 1903', *New York History*, 78 (1997), pp. 273–308; N. Richardson, *Typhoid in Uppingham* (London: Pickering & Chatto, 2008); A. Hardy, 'Exorcising Molly Malone: Typhoid and Shellfish Consumption in Urban Britain 1860–1960', *History Workshop Journal*, 55 (2003), pp. 73–90; D. F. Smith and H. L. Diack with T. H. Pennington and E. M. Russell, *Food Poisoning, Policy and Politics: Corned Beef and Typhoid in Britain in the 1960s* (Woodbridge: The Boydell Press, 2005).

3. E. O. Price, 'The Bangor Typhoid Epidemic of 1882', MD thesis, 1891, reprinted in *Caernarvonshire Historical Society Transactions*, 26 (1965), pp. 157–68, on pp. 158, 167; North Riding County Council, *Annual Report of the County Medical Officer of Health for the Year 1932*, pp. 1, 34.

4. Tomes, 'The Private Side', p. 519.

5. M. Humphries, 'Typhoid and its Carriers', in K. F. Kiple (ed.), *Plague, Pox and Pestilence: Disease in History* (London: Weidenfeld and Nicolson, 1997), pp. 14–19, on pp. 16–17; LeBaron and Taylor, 'Typhoid Fever', pp. 1073–4; Hardy, 'On the Cusp', p. 331.

6. Humphries, 'Typhoid', p. 14.

7. P. Wald, *Contagious: Cultures, Carriers, and the Outbreak Narrative* (Durham, NC: Duke University Press, 2008), p. 82.

8. W. Budd, *Typhoid Fever, its Nature, Mode of Spreading and Prevention* (New York, 1873, reprinted 1931), p. 1, cited in Stevenson, 'Exemplary Disease', p. 178.

9. A. Briggs, 'Cholera and Society in the Nineteenth Century', in A. Briggs, *The Collected Essays of Asa Briggs, Volume II: Images, Problems, Standpoints, Forecasts* (Brighton: The Harvester Press, 1985), pp. 153–76, on p. 154.

10. A. Wohl, *Endangered Lives: Public Health in Victorian Britain* (London: Methuen and Co., 1984), p. 128; R. Woods, 'Mortality and Sanitary Conditions in Late Nineteenth-Century Birmingham', in R. Woods and J. Woodward (eds), *Urban Disease and Mortality in Nineteenth Century England* (London and New York: Batsford Academic and Educational, and St. Martin's Press, 1984), pp. 176–202, on p. 197; Worboys, *Spreading Germs*, p. 132; C. Murchison, *A Treatise on the Continued Fevers of Great Britain* (London: 1884), p. 441, cited in Hardy, 'On the Cusp', p. 332.

11. Mallon was aware of bacteriology and the carrier concept, and had her own private laboratory tests done which did not show the typhoid bacillus, contrary to those conducted by New York City. See J. A. Mendelsohn, '"Typhoid Mary" Strikes Again: The Social and the Scientific in the Making of Modern Public Health', *Isis*, 86 (1995), pp. 268–77, on p. 273. See also Wald, *Contagious*; J. W. Leavitt, *Typhoid Mary: Captive to the Public's Health* (Boston, MA: Beacon Press, 1996); Richardson, *Typhoid in Uppingham*; Prescott, 'Sending their Sons into Danger'; Smith and Diack, *Food Poisoning*; Hardy, 'Exorcising Molly Malone'; Hardy, 'Scientific Strategy'.

12. 'The Health of Bangor', *Carnarvon and Denbigh Herald*, 26 August 1882, p. 5.

13. Ibid.

14. Price, 'Bangor Typhoid Epidemic of 1882', p. 167.

15. Worboys, *Spreading Germs*; Tomes, 'The Private Side'.

16. Ibid., p. 518.

17. Stevenson, 'Science Down the Drain', pp. 1–14.

18. Tomes, 'The Private Side', pp. 509–11, 519, 524.

19. Weekly Meeting, Report on the Drainage of Addenbrooke's Hospital, 24 January 1883, AHA, AHM 31.

20. Weekly Meeting, 21 February 1883, AHA, AHM 31; Weekly Meeting, 15 September 1886, AHA, AHM 33; Weekly Meeting, 9 December 1887, AHA, AHM 33.

21. 'The Bangor Local Board', *Carnarvon and Denbigh Herald*, 12 August 1882, p. 9.

22. 'The Health of Bangor', *Carnarvon and Denbigh Herald*, 19 August 1882, p. 8.

23. 'Bangor', *Carnarvon and Denbigh Herald*, 30 September 1882, p. 5.

24. 'Bangor', *Carnarvon and Denbigh Herald*, 14 October 1882, p. 8.

25. See examples, 'The Bangor Epidemic', *Carnarvon and Denbigh Herald*, 14 October 1882, p. 6, including references to letters in the *Lancet*; 'The Bangor Local Board and Dr Rees', *Carnarvon and Denbigh Herald*, 28 October 1882, p. 6.

26. 'Critical utterances at Bangor', *Carnarvon and Denbigh Herald*, 28 October 1882, p. 4.

27. A collection of five letters from Florence Nightingale to William Rathbone, MP for Caernarfonshire, concerning the typhoid epidemic at Bangor in 1882, University of Wales, Bangor, Department of Archives and Manuscripts, 13 October 1882, MSS 37619; 22 September 1882, MSS 37616.

28. Letters from Florence Nightingale to William Rathbone, University of Wales, Bangor, Department of Archives and Manuscripts, 22 September 1882, MSS 37617.

29. C. E. Rosenberg, 'Florence Nightingale on Contagion: The Hospital as Moral Universe', in C. E. Rosenberg (ed.), *Explaining Epidemics and Other Studies in the History of Medicine* (New York: Cambridge University Press, 1992), pp. 116–36, esp. pp. 92, 100–4.

30. L. McDonald, 'Mythologizing and De-Mythologizing', in S. Nelson and A. M. Rafferty (eds), *Notes on Nightingale* (Ithaca, NY: Cornell University Press, 2010), pp. 91–114, on pp. 99–100.

31. For example, 'The Bangor Epidemic and Benevolent Services' and 'Typhoid Fever', *Carnarvon and Denbigh Herald*, 23 September 1882, p. 4.

32. 'The Bangor Epidemic and Benevolent Services', *Carnarvon and Denbigh Herald*, 30 September 1882, p. 4.

33. 'Bangor', *Carnarvon and Denbigh Herald*, 4 November 1882, p. 5.

34. McGrew, *Russia and the Cholera*; Snowden, *Naples in the Time of Cholera*.

35. See Glanogwen National Girls' School 1881–1911 Log Book, 7 August 1882, 21–5 August 1882, 26 October 1882, 30 October 1882, Caernarfon Record Office (hereafter CaRO), XES1/14/1, pp. 17, 20, 21; Upper Bangor National School Log Book 1892–89, 9 October 1882, CaRO, XES1/6/1, p. 225; National School, Bangor, Boys' Department Log Book 1876–1906, CaRO, XES1/15/1, p. 105.

36. In the Bangor National Girls School Log Book it was recorded that the Thanksgiving service for Harvest was held 'in all the Chapels in the town'. See Bangor National Girls School 1876–1914 Log Book, 9 October 1882, CaRO, XES1/16/1, p. 115. See also Bangor National Infants School 1870–1960 Log Book, 9 October 1882, CaRO, XES1/14/1, p. 209; Upper Bangor National School Log Book 1892–89, CaRO, XES1/6/1, p. 224; National School, Bangor, Boys' Department Log Book 1876–1906, 9 October 1882, CaRO, XES1/15/1, p. 105; 10 October 1882 for further church services which were detracting from school attendance, p. 105. Also see 'Bangor', *Carnarvon and Denbigh Herald*, 14 October 1882, p. 8.

37. The Board School, Garth Bangor Log Book March 1881–April 1892 Log Book, 15 August 1882, CaRO, XES1/18/2, pp. 97–8.

38. National School, Bangor, Boys' Department 1876–1906 Log Book, 14 August 1882 and 25 September 1882, CaRO, XES1/15/1, p. 104.

39. Letter from Mr S., 9 October 1858; Letter from Rev. B., 22 October 1858, Marlborough College Archives (hereafter MCA), Unsorted Letters, Bursar Tomkinson, 1858.

40. See J. Wolffe, 'Religion and Secularization', in P. Johnson (ed.), *Twentieth-Century Britain: Economic, Social and Cultural Change* (London and New York, Longman, 1994), pp. 427–41, on pp. 428–9; Conversation with Marlborough College Archivist, Terry Rogers.
41. Rosenberg, *The Cholera Years*, pp. 40–54, 213–14.
42. Letter from Mrs M, 6 October 1858, MCA, Unsorted Letters, Bursar Tomkinson, 1858.
43. Letter from Mrs M., 9 October 1858, MCA, Unsorted Letters, Bursar Tomkinson, 1858; Circular from G. G. Bradley, Master, to parents, 7 October 1858, MCA, Medical Box 331.
44. Letter from Mr M., 9 October 1858, MCA, Unsorted Letters, Bursar Tomkinson, 1858.
45. Richardson, *Typhoid in Uppingham*, p. 1.
46. 'Diphtheria', *Malburian*, 25 (29 November 1890), p. 180, including extract from the *British Medical Journal*.
47. Annual General Meeting of the Life Governors of Marlborough College, held at the Charing Cross Hotel, 27 May 1891, MCA, Marlborough College Minutes of General Meetings, 1845–93.
48. Letter from Rev. G. C. Bell, Master, to parents, 28 October 1890, reprinted in 'Diphtheria', *Malburian*, 25 (29 November 1890), p. 180.
49. Letter from Rev. G. C. Bell, Master, to parents, 6 November 1890, reprinted in 'Diphtheria', *Malburian*, 25 (29 November 1890), p. 182.
50. Letter from Mr D. to Mr Richardson, 28 October 1890; Letter from Mr A., 9 November 1890, MCA, Letters from parents to the Bursar, Bursar Thomas 1889–90, box 225.
51. Letter from Rev. F. to Thomas, 27 October 1890, MCA, Letters from parents to the Bursar, Bursar Thomas 1889–90, box 225.
52. Richardson, 'Typhoid in Uppingham', p. 10.
53. Prescott, 'Sending their Sons into Danger', p. 308.
54. Local Government Board, *Twenty-Seventh Annual Report of the Local Government Board, 1897–98* (London: HMSO, 1898), appendix B, p. 70.
55. 'Epidemic', *South Eastern Gazette*, 21 September 1897, p. 5.
56. 'The Health of Maidstone', *Kent Messenger*, 25 September 1897, p. 9.
57. 'Letters', *South Eastern Gazette*, 12 October 1897, p. 5.
58. 'Maidstone Epidemic', *South Eastern Gazette*, 1 January 1898, p. 2.
59. 'Letter to the Editor', *Kent Messenger*, 25 September 1898, p. 10.
60. 'Enteric Fever in Maidstone', *Kent Messenger*, 25 September 1897, p. 10.
61. 'Letter to the Editor', *Kent Messenger*, 16 October 1897, p. 3.
62. 'Maidstone's Plague', *South Eastern Gazette*, 2 October 1897, p. 2.
63. 'Enteric Fever in Maidstone', *Kent Messenger*, 25 September 1897, p. 10.
64. Hamlin, *A Science of Impurity*, p. 264. See also Hardy, 'On the Cusp', p. 333.
65. 'Letters', *South Eastern Gazette*, 5 October 1897, p. 5.
66. 'Proposed Conference with the Water Company', 1 January 1898, p. 2.
67. 'Maidstone Epidemic', *South Eastern Gazette*, 4 January 1898, p. 5.
68. 'Epidemic', *South Eastern Gazette*, 21 September 1897, p. 5; 'The Typhoid Scourge at Maidstone', *South Eastern Gazette*, 28 September 1897, p. 5; 'The Health of Maidstone', *Kent Messenger*, 25 September 1897, p. 9; 'Account of Investigation by the Special Correspondent', *Kent Messenger*, 2 October 1897, p. 11.
69. 'Letters', *Kent Messenger*, 25 September 1897, p. 10.
70. 'Letter to the Editor', *Kent Messenger*, 2 October 1897, p. 9.
71. 'The Clergy on the Situation', *Kent Messenger*, 2 October 1897, p. 11.
72. 'Inquiry Over', *Kent Messenger*, 12 February 1898, p. 5.

73. 'Questioning Scientists', *Kent Messenger*, 19 March 1898, p. 6.
74. 'Letter to the Editor', *Kent Messenger*, 25 September 1897, p. 10.
75. 'Maidstone Epidemic', *South Eastern Gazette*, 1 January 1898, p. 2.
76. 'Enteric Fever in Maidstone', *Kent Messenger*, 25 September 1897, p. 10.
77. 'Letter to the Editor', *Kent Messenger*, 2 October 1897, p. 11.
78. Ibid.
79. 'Letters', *South Eastern Gazette*, letters dated 5 October 1897, p. 5.
80. 'Maidstone UDC Meeting', *Kent Messenger*, 9 October 1897, p. 8.
81. 'Letter to the Editor', *Kent Messenger*, 9 October 1897, p. 9.
82. 'UDC Meeting', *Kent Messenger*, 16 October 1897, p. 8.
83. 'Letter to the Editor', *Kent Messenger*, 22 January 1898, p. 5.
84. 'Inquiry Over', *Kent Messenger*, 12 February 1898, p. 5.
85. 'Private Meeting', *Kent Messenger*, 5 March 1898, p. 5.
86. '£3,000 Grant', *Kent Messenger*, 16 April 1897.
87. Taylor and Trentmann, 'Liquid Politics'.
88. 'Netley', *South Eastern Gazette*, 9 October 1897, p. 2.
89. J. R. Matthews, *Quantification and the Quest for Medical Certainty* (Princeton, NJ: Princeton University Press, 1995), p. 98.
90. R. J. Reece, 'Appendix A. No. 8. Report on the Epidemic of Enteric Fever in the City of Lincoln, 1904–5', in Local Government Board, *Supplement to the Thirty-fifth Annual Report of the Local Government Board, 1905–06, containing the Report of the Medical Officer for 1905–06*, pp. 81–145 (London: HMSO, 1907), vol. 17, p. 96.
91. H. Velten, *Milk: A Global History* (London: Reaktion, 2010), p. 79.
92. Valenze, *Milk*, p. 154, 177, 211–13.
93. Tomes, 'The Private Side', p. 512.
94. Wald, *Contagious*, p. 16.
95. R. McKay, 'Imagining "Patient Zero": Sexuality, Blame, and the Origins of the North American AIDS Epidemic' (PhD thesis, University of Oxford, 2010), p. 64.
96. A. Ledingham and J. C. G. Ledingham, 'Typhoid Carriers', *British Medical Journal* (4 January 1908), pp. 15–17.
97. Mendelsohn, 'Cultures of Bacteriology', pp. 633–79, 694.
98. McKay, 'Imagining "Patient Zero"', pp. 66–8, citing D. S. Davies and I. Walker Hall, 'A Discussion on the Etiology and Epidemiology of Typhoid (Enteric) Fever: Typhoid Carriers, with an Account of Two Institution Outbreaks Traced to the Same "Carrier"', *Proceedings of the Royal Society of Medicine*, 1 (1907–8), pp. 175–91.
99. Leavitt, *Typhoid Mary*; Wald, *Contagious*, p. 69.
100. Mendelsohn, '"Typhoid Mary" Strikes Again', p. 271.
101. Ibid., p. 276–7; Leavitt, *Typhoid Mary*, p. 626.
102. Wald, *Contagious*, pp. 73–4.
103. Tomes, 'The Private Side', p. 513.
104. 'Enteric Fever at Eccles', *Eccles and Patricroft Journal*, 25 November 1910, p. 10.
105. '7 further cases', *Eccles and Patricroft Journal*, 2 December 1910, p. 10.
106. 'Eccles Town Council Meeting', *Eccles and Patricroft Journal*, 9 December 1910, p. 9.
107. 'Letter to the Editor', *Eccles and Patricroft Journal*, 30 December 1910, p. 9; appeal in 'Appeal for Money', *Eccles and Patricroft Journal*, 23 December 1910, p. 10.
108. For example, 'Letters to the Editor', *Eccles and Patricroft Journal*, 2 December 1910, p. 9.
109. Local Government Board, *Annual Report of the Local Government Board, Supplement Containing the Report of the Medical Officer for 1910–11* (London: HMSO, 1912), pp. 21–3.

110. Ibid., p. 212.
111. 'Enteric Fever at Eccles', *Eccles and Patricroft Journal*, 25 November 1910, p. 10; 'Eccles Town Council Meeting', 9 December 1910, p. 9; *Annual Report of the Local Government Board, Supplement for 1910–11*, pp. xi–xii.
112. *Annual Report of the Local Government Board, Supplement for 1910–11*, p. 28.
113. Ibid., pp. 28–9.
114. '7 Further Cases'.
115. 'Enteric Epidemic', *Rotherham Advertiser*, 12 November 1921, p. 12.
116. R. M. F. Picken, 'Age and Sex Incidence of Milk Borne Epidemics', *British Medical Journal* (27 June 1936), pp. 1291–5.
117. 'The Typhoid Outbreak', *Gravesend and Dartford Reporter*, 21 May 1927, p. 7; 'Council and the Epidemic', *Gravesend and Dartford Reporter*, 28 May 1927, p. 3; Question from Colonel Day to Mr Chamberlain, Hansard, HC 16 June 1927.
118. 'Council and the Epidemic', *Gravesend and Dartford Reporter*, 28 May 1927, p. 3.
119. 'Typhoid in Our Midst', *Gravesend and Dartford Reporter*, 28 May 1927, p. 10.
120. Ministry of Health, *Tenth Annual Report of the Ministry of Health, 1928–1929* (London: HMSO, 1929), p. 38.
121. 'Para-Typhoid Outbreak', *The Times*, 11 August 1911, p. 7.
122. 'Scarlet Pimpernel Germ Drama', *Evening Standard*, 8 August 1928, p. 10.
123. 'Editorial', *Evening Standard*, 13 August 1928, p. 6.
124. 'Paratyphoid in London', *The Times*, 4 August 1928, p. 9.
125. 'More Cases of Paratyphoid in London'. *The Times*, 15 February 1928, p. 15.
126. 'Paratyphoid Fever in London', *British Medical Journal* (11 May 1929), pp. 871–2; 'Para-Typhoid', *The Times*, 10 August 1928, p. 14; 'Food Preservation', *The Times*, 9 October 1928, p. 17; Valenze, *Milk*, p. 211.
127. 'Letter to the Editor', *The Times*, 6 August 1928, p. 6.
128. 'Paratyphoid Fever in London', pp. 871–2.
129. North Riding of Yorkshire County Council, *Annual Report of the County Medical Officer of Health for the Year 1932* (York: North Riding County Council, 1933), p. 34.
130. 'The Typhoid Report – and After', no newspaper title, 17 March 1933, Press cuttings, NYRO TRF. The newspaper title, page and date are often not preserved for these newspaper cuttings.
131. 'Cost of Typhoid Epidemic', no newspaper title, 2 June 1933, Press cuttings, NYRO TRF.
132. 'Letter to the Editor', *Malton Messenger*, letter dated 2 January 1933, Press cuttings, NYRO TRF.
133. 'Letters to the Editor', no newspaper title, letter dated 2 January 1933; no newspaper title, letter dated 4 January 1933, Press cuttings, NYRO TRF. For more examples of complaints about water see 'Letters to the Editor', *Malton Messenger*, 31 December 1932; no newspaper title, 2 December 1932, Press cuttings, NYRO ZPB.
134. 'The Typhoid Outbreak', no newspaper title, 28 October 1932, reports that the water analysis results had still not arrived, Press cuttings, NYRO TRF; 'Latest News', *Malton Messenger*, 12 November 1932, File on water, NYRO TRF.
135. 'Heavy Toll of Typhoid Epidemic', *Malton Messenger*, 5 November 1932; 'Latest News', *Malton Messenger*, 12 November 1932, Press cuttings, NYRO TRF.
136. 'Emergency Meeting: The Water Analysis', *Malton Messenger*, 5 November 1932, Press cuttings, NYRO TRF. See also D. J. Dawson and D. P. Sartory, 'Microbiological Safety of Water', *British Medical Bulletin*, 56 (2000), pp. 74–83.

137. 'Malton Water Surprise', *Yorkshire Herald*, 22 December 1932, Press cuttings, NYRO TRF.

138. Quote from 'Letter to the Editor', *Malton Messenger*, letter dated 2 January 1933; 'Malton's Water', no newspaper title, 12 May 1933, Press cuttings, NYRO TRF; Malton, *Report of the Medical Officer of Health for the Year 1932*, p. 4. See also Prescott, 'Sending their Sons into Danger', p. 303, for the same discussion in Ithaca, New York.

139. 'Malton Water Surprise', *Yorkshire Herald*, 22 December 1932, NYRO TRF, box 1.

140. 'Proposed Water Scheme for Malton: Public Enquiry', Unlabelled newspaper, 22 July 1933, File on water, NYRO TRF.

141. 'Complaints from the MOH', *Yorkshire Gazette*, 31 March 1933, NYRO TRF, box 3.

142. 'Sixty Typhoid Cases', *Yorkshire Gazette*, 28 October 1933, Press cuttings, NYRO TRF.

143. 'Letter to the Editor. The Typhoid Disaster', *Malton Messenger*, letter dated 31 December 1932, Press cuttings, NYRO TRF.

144. 'Letter to the Editor', *Malton Messenger*, undated, letter written 2 January 1933, NYRO TRF, box 3.

145. 'Letter to the Editor', no newspaper title, letter dated 21 January 1933, File on water, NYRO TRF.

146. Letter to the Editor', *Malton Messenger*, letter dated 21 December 1932, Press cuttings, NYRO TRF.

147. Letters between Mrs A. and Mr E. and the Relief Committee, February to May 1933, NYRO TRF, box 1.

148. Letters between Mr S. and the Relief Committee, 21 and 22 February 1933, NYRO TRF, box 1.

149. Letter, February 1934, File of Administrative Papers, NYRO TRF, box 1.

150. Notes on separate pieces of paper regarding instances where wealthier people, or those bringing in no income or a very low income, requested compensation; File – Malton and District Relief Fund, typewritten sheets about the Relief Committee, NYRO TRF, box 1.

151. Letters between Mrs M., the Darlington Branch and Ronald Elston, January 1933; Letters between Mr R. and the Relief Committee, February 1933, NYRO TRF, box 1.

152. Letters between Relief Committee and Mrs C. and Miss D., January to February 1933, p. 14, NYRO TRF, box 1.

153. Letters between Inspector of Police, Shipton, and Relief Committee, February 1933, NYRO TRF, box 1.

154. Letter from Captain Gibson to Kit, 21 March 1933, NYRO TRF, box 1; Malton and District Relief Fund Minute Book, 10 April 1934, NYRO TRF.

155. Malton and District Relief Fund Minute Book, 27 May 1933, NYRO TRF.

156. 'Dr L. C. Walker on the Epidemic', *Yorkshire Gazette*, 18 November 1932, File of Administrative Papers, NYRO TRF, box 3. However, according to one newspaper correspondent, private warnings were still issued before the public ones. 'Letter to the Editor', no newspaper title, letter dated 21 January 1933, Press cuttings, NYRO TRF.

157. For example, leaflets and notices were distributed by the council, and there was a poster distributed which was approved by the council which gave instructions about how typhoid was spread and that shopping was safe. See 'Letter to the Editor', letter dated 23 December 1932, Press cuttings, NYRO TRF.

158. Letters from J. G. Kinimouth, H. de Mirimonde, N. C. Forsyth, L. C. Walker, 2–6 March 1933, NYRO TRF, box 1.

159. Letter from Malton and Norton Unemployed Committee Fund to Typhoid Relief Committee, 24 March 1933, NYRO TRF, box 1; 'A Princely Gesture', no newspaper title, 12 May 1933, Press cuttings, NYRO TRF.

160. 'Letters to the Editor', *Malton Messenger*, 19 November 1932; 6 January 1933, Press cuttings, NYRO TRF.

161. 'Epidemic and Trade', *Yorkshire Gazette*, 18 November 1932, Press cuttings, NYRO TRF.

162. 'A Princely Gesture', no newspaper title, 12 May 1933, Press cuttings, NYRO TRF.

163. Letter from Relief Committee to North Riding Lieutenancy, 6 January 1933, NYRO TRF, box 1.

164. Report of the Typhoid Relief Committee (efforts of the Committee), File of Administrative Papers, NYRO TRF; Letter from Violet Chomley to Captain Gibson, 9 December 1932, saying a dance is being held to raise money, NYRO TRF, box 1; 'Our Relief Fund Now Totals £1,388', *Yorkshire Herald*, 12 December 1932 (boxing gloves and nativity play); 'Our Malton Fund £2,180', *Yorkshire Herald*, 24 December 1932 (football match); 'A Princely Gesture', Unlabelled newspaper, 12 May 1933, Press cuttings, NYRO TRF.

165. 'Letter to the Editor', no newspaper title, letter dated 21 January 1933, Press cuttings, NYRO TRF.

166. 'Letter to the Editor', *Malton Messenger*, letter dated 31 December 1932, Press cuttings, NYRO TRF.

167. M. Graham, *The Typhoid Epidemic in Bournemouth, Poole and Christhurch 1936* (Christchurch: Bournemouth Local Studies Publications, 1997); Smith and Diack, *Food Poisoning*, p. 11.

168. Graham, *Typhoid Epidemic*, p. 7.

169. A. S. MacNalty, 'Prefatory Note by the Chief Medical Officer', in Shaw, 'Report on an Outbreak of Enteric Fever', pp. 3–4.

170. Graham, *Typhoid Epidemic*, pp. 1, 6–7.

171. Ibid., pp. 8, 27.

172. Ibid., pp. 12–14.

173. MacNalty, 'Prefatory Note by the Chief Medical Officer', pp. 3–4.

174. W. Coleman, *Yellow Fever in the North: The Methods of Early Epidemiology* (Madison, WI: The University of Wisconsin Press, 1987), p. 194.

175. Ibid., pp. xvi, 3, 82, 119.

176. Hamlin, *A Science of Impurity*, pp. 260, 289; Hardy, 'On the Cusp', pp. 336–7.

177. Hamlin, *A Science of Impurity*, p. 270; See also Hardy, 'On the Cusp', p. 328.

178. Hamlin, *A Science of Impurity*, pp. 271–2, 293; A. Hardy, 'Methods of Outbreak Investigation in the "Era of Bacteriology" 1880–1920', *Sozial Präventivmed*, 46 (2001), pp. 355–60, on p. 357.

179. Hardy, 'On the Cusp', p. 334.

180. Ibid., p. 343.

181. A successful community claim for compensation had been made in the US after an epidemic in Olean, New York in 1928, amounting to $425,000, so the Croydon epidemic was not the first successful case internationally. See Hardy, 'Scientific Strategy', p. 18.

6 The 'Typhoid Trials': Compensation Culture in 1930s Croydon

1.   Ministry of Health, *Report on a Public Local Inquiry into an Outbreak of Typhoid Fever at Croydon in October and November 1937* (London: HMSO, 1938), p. 4.

2. Finance Committee Meeting, 30 October 1939, Finance Committee Minute Book, 1 November 1938 to 31 October 1939, Croydon Local Studies Library and Archives Service (hereafter CAS), County Borough of Croydon (hereafter CBC).

3. F. Honigsbaum, *The Division in British Medicine: A History of the Separation of General Practice from Hospital Care 1911–1968* (New York: St Martin's Press, 1979), pp. 148, 192–4, 64–6, 206; C. Webster, *The National Health Service: A Political History* (Oxford: Oxford University Press, 2002), p. 5; J. Gent, *Croydon Past* (Chichester: Phillimore and Co. Ltd, 2002), p. 109.

4. Inquiry into the Outbreak of Typhoid Fever in October and November 1937, Minutes of Proceedings, Town Clerk, Croydon, CAS, CBC, fs70 (614.4) COR (hereafter Typhoid Inquiry), p. 2.

5. Ministry of Health, *Report on a Public Local Inquiry*, p. 3.

6. The local council was known as Croydon Corporation, often referred to as the Corporation.

7. M. Lyons, Typhoid Inquiry, 6 January 1938, pp. 741–3.

8. Letter from C. Rimington to O. Holden, 29 October 1937, read by M. Lyons, 6 January 1938, Typhoid Inquiry, p. 742.

9. Rimington, Typhoid Inquiry, 6 January 1938, pp. 743–4, 752–4.

10. R. Moss, Typhoid Inquiry, 5 January 1938, p. 683.

11. Rimington, Typhoid Inquiry, 6 January 1938, p. 745.

12. Moss, Typhoid Inquiry, 5 January 1938, p. 683–4.

13. M. K. Humphreys, Typhoid Inquiry, 6 January 1938, pp. 740–1.

14. Rimington, Typhoid Inquiry, 6 January 1938, p. 745.

15. Holden, Typhoid Inquiry, 22 December 1937, pp. 187–8.

16. Rimington, Typhoid Inquiry, 6 January 1938, p. 746–7.

17. Ibid., pp. 744–5.

18. Moss, Typhoid Inquiry, 5 January 1938, p. 684; C. Green, Typhoid Inquiry, 10 January 1938, p. 861.

19. Ibid.

20. Moss, Typhoid Inquiry, 5 January 1938, pp. 685–6.

21. Ibid., p. 686, 689.

22. G. Howard, Typhoid Inquiry, 5 January 1938, p. 692.

23. W. Monckton, Typhoid Inquiry, 6 January 1938, p. 750.

24. Rimington, Typhoid Inquiry, 6 January 1938, p. 753; Ministry of Health, *Report on a Public Local Inquiry*, p. 9.

25. Ibid., p. 14.

26. C. Chapin, 'The End of the Filth Theory of Disease', *Popular Science Monthly*, 60 (1902), pp. 234–9, cited in Tomes, *Gospel of Germs*, p. 108.

27. See p. 136.

28. N. Rogers, 'Germs with Legs: Flies, Disease, and the New Public Health', *Bulletin of the History of Medicine*, 63 (1989), pp. 599–617, on pp. 599–601.

29. A. Hamilton, 'The Fly as a Carrier of Typhoid: An Inquiry into the Part Played by the Common House Fly in the Recent Epidemic of Typhoid Fever in Chicago', *Journal of the American Medical Association*, 40 (1903), pp. 576–83, cited in Rogers, 'Germs with Legs', pp. 599–617, on p. 607.

30. J. T. C. Nash, 'House Flies as Carriers of Disease', *Journal of Hygiene*, 9 (1909), p. 149, cited in Rogers, 'Germs with Legs', p. 606.

31. J. F. M. Clark, *Bugs and the Victorians* (New Haven, CT, and London: Yale University Press, 2009), p. 222.
32. Ibid., pp. 222–4.
33. Ibid., pp. 227–8.
34. Ibid., p. 11.
35. Ibid., pp. 217–8.
36. 'Medical Association of Georgia Report', *Journal of the American Medical Association*, 52 (1909), p. 2019, cited in Rogers, 'Germs with Legs', p. 605.
37. Clark, *Bugs and the Victorians*, p. 219.
38. Ministry of Health, *Report on a Public Local Inquiry*, p. 16.
39. Ibid., pp. 16–17.
40. Ibid., p. 14. In the test case he is referred to as the Ministry's specialist on typhoid. Test Action, A. and P. R. Read v. Croydon Corporation, November 1938, CAS, fs70 (614.4) CRO, Wallington, 10 November 1938, p. 3.
41. Ministry of Health, *Report on a Public Local Inquiry*, p. 4.
42. Ibid., p. 12; H. J. Wallington, Test Action, 11 November 1938, p. 59.
43. Ministry of Health, *Report on a Public Local Inquiry*, pp. 15–16.
44. Ibid., p. 11. E. Conybeare, Typhoid Inquiry, 7 January 1938, pp. 792–4.
45. Moss, Typhoid Inquiry, 5 January 1938, pp. 690–1.
46. Rimington, Typhoid Inquiry, 6 January 1938, pp. 746, 756, 747.
47. G. Lewin, Typhoid Inquiry, 6 January 1938, p. 760.
48. Rimington, Typhoid Inquiry, 6 January 1938, p. 759; *News Chronicle*, 13 November 1937 in Sir W. Monckton, K. C., 'Newspaper Cuttings Relating to the Typhoid Outbreak in Croydon', vol. 2, Nov. 26th–Dec. 4th', CAS, fs70 (614.4) CRO (hereafter Newspaper cuttings).
49. Rimington, Typhoid Inquiry, 6 January 1938, pp. 758–9.
50. Monckton, Typhoid Inquiry, 21 December 1937, p. 97.
51. Lyons, Typhoid Inquiry, 3 January 1938, p. 565.
52. R. Jackson, Typhoid Inquiry, 4 January 1938, pp. 585–6.
53. Records of the progress of cases at Whitgift of Typhoid victims of the outbreak in Oct. Nov. 1937, Statement for 20 November 1937, Whiftgift School Archives, Croydon, SM21/3g.
54. Commons Sitting, 18 November 1937, Hansard, HC 329.
55. Commons Sitting, 25 November 1937, Hansard, HC 329.
56. Commons Sitting, 30 November 1937, Hansard, HC 329.
57. Written answers (Commons), 21 February, 1938, Hansard, HC 329; Written answers (Commons), Ellen Wilkinson, Jarrow, Communist Party and Labour Party, question to Sir Kingsley Wood, 24 February 1938, Hansard, HC 329; Written answers (Commons) Thomas Groves, West Ham Stratford, Labour Party Whip, to Sir Kingsley Wood, 2 March 1938, Hansard, HC 329; Commons sitting, Local Authorities (Enabling) Bill, Thomas Levy, Elland, Conservative Party, 4 March 1938, Hansard, HC 329.
58. W. H. Wilcox, Typhoid Inquiry, 6 January 1938, p. 723. For problems in detecting typhoid in water, see Hamlin, *A Science of Impurity*, pp. 289–91.
59. Ministry of Health, *Report on a Public Local Inquiry*, p. 4; Hamlin, *A Science of Impurity*, pp. 300–1.
60. Typhoid Inquiry, 6 December 1937, p. 14.
61. 'Typhoid Terror Mystery', *Sunday Times*, 21 November 1937, Croydon press cuttings, vol. 1, S70(614.4) CRO (hereafter Croydon press cuttings).

62. Holden, Typhoid Inquiry, 22 December 1937, p. 204; Holden, Typhoid Inquiry, 23 December 1937, pp. 249–50. For interviews with workers (one of which was the carrier, but was interviewed anonymously as one of the workers), see 4 January 1938, pp. 577–84. See p. 584 for explanation of anonymity; Ministry of Health, *Report on a Public Local Inquiry*, p. 3.

63. Wallington, Test Action, 11 November 1938, p. 61.

64. Wallington and H. Brandram-Jones, Test Action, 10 November 1938, p. 39. See also Holden, Typhoid Inquiry, 22 December 1937, p. 204; Holden, Typhoid Inquiry, 23 December 1937, pp. 249–50. Brandram-Jones moved to Brighton for a new job during October.

65. Wallington and Brandram-Jones, Test Action, 17 November 1938, p. 306.

66. W. M. Scott and Murphy, Typhoid Inquiry, 7 January 1938, p. 806.

67. Foster, *History of Medical Bacteriology*, p. 194.

68. Hardy, 'Scientific Strategy', pp. 27, 29; Shaw, 'Report on an Outbreak of Enteric Fever', p. 16.

69. J. S. K. Boyd, 'Bacteriophage-Typing and Epidemiological Problems', *British Medical Journal* (27 September 1952), p. 4786.

70. John Eyre, Test Action, 11 November 1938, pp. 90–5.

71. Written answers (Commons), Wilfred Roberts, MP, Liberal and Labour Party, Cumberland Northern, written question to Sir Kingsley Wood, Minister of Health, Conservative Party, 25 November 1937, Hansard, HC 329.

72. J. Fenton and C. Hay, 'Prophylactic Use of Anti-Typhoid Serum in a Localized Outbreak', *British Medical Journal* (21 May 1938), pp. 1090–1.

73. 'Typhoid Victims now 127', *Observer*, 21 November 1937, Newspaper cuttings, vol. 1, p. 78.

74. Holden and Lyons, Typhoid Inquiry, 28 December 1937, pp. 309–10.

75. 'More Typhoid', *The Times*, 22 November 1937, p. 12.

76. 'Letter to the Editor', *The Times*, 25 November 1937, p. 10

77. 'Typhoid Fever at Croydon', *British Medical Journal* (20 November 1937), in Newspaper cuttings, vol. 1, p. 67.

78. S. Watson Smith, 'Correspondence', *British Medical Journal* (4 December 1937), p. 1139; E. W. Goodall, 'Correspondence', *British Medical Journal* (11 December 1937), p. 1197.

79. S. Watson Smith, 'Correspondence', *British Medical Journal* (4 December 1937), p. 1139.

80. R. Taylor, 'Correspondence', *British Medical Journal* (4 December 1937), pp. 1139–40; R. Taylor, 'Typhoid Fever in the Basque Refugee Camp', *British Medical Journal* (16 October 1937), pp. 760–1.

81. Major Greenwood, 'Correspondence', *British Medical Journal* (11 December 1937), pp. 1197.

82. 'Immunization against Typhoid', *British Medical Journal* (4 December 1937), pp. 1128–9.

83. W. Wilcox, 'Typhoid Fever: Its Clinical Aspects', *British Medical Journal* (21 May 1938), pp. 1089–90. Willcox attended the trial and voiced the idea that local areas should have local committees where information could be pooled and distributed. See 'The Croydon Typhoid Inquiry', *British Medical Journal* (19 February 1938), p. 407.

84. W. H. Wilcox, Typhoid Inquiry, 6 January 1938, p. 726.

85. Written answers (Commons), Thomas Groves, MP, Labour Party Whip, West Ham Stratford, question to Sir Kingsley Wood, 2 March 1938, Hansard, HC 329.

86. 'Typhoid Terror Mystery', *Sunday Times*, 21 November 1937, Croydon press cuttings, vol. 1.

87. Mr Harris to Holden, Typhoid Inquiry, 6 January 1938, pp. 700–1.

88. J. V. Pincus, 'Correspondence', *British Medical Journal* (4 December 1937), p. 1140.

89. E. M. Fraenkel, 'Immunization against Typhoid', *British Medical Journal* (1 January 1938), p. 41.

90. S. Watson Smith, 'Correspondence', *British Medical Journal* (4 December 1937), p. 1139; 'Immunization against Typhoid', *British Medical Journal* (4 December 1937), pp. 1128–9; W. Wilcox, 'Typhoid Fever: Its Clinical Aspects', *British Medical Journal* (21 May 1938), p. 1089; E. W. Goodall, 'Correspondence', *British Medical Journal* (11 December 1937), p. 1197.

91. J. Fenton and C. Hay, 'Prophylactic Use of Anti-Typhoid Serum in a Localized Outbreak', *British Medical Journal* (21 May 1938), pp. 1090–1.

92. 'Typhoid Serum', *Croydon Times*, 24 November 1937, Newspaper cuttings, vol. 1, p. 139.

93. K. V. Subbarao, 'Correspondence', *British Medical Journal* (17 September 1938), p. 638.

94. Wilcox, 'Typhoid Fever', pp. 1085–90.

95. Ibid., pp. 1087–9.

96. Ministry of Health, *Report on a Public Local Inquiry*, p. 11.

97. Mendelsohn, '"Typhoid Mary" Strikes Again', p. 270.

98. However, in the 1930s, in the same decade as the Croydon decision of how to tackle carriers, New York City abandoned its practice of testing food handlers as they were given certificates for a period of time, and it could not be guaranteed that they would not become infected with typhoid before they were next tested. See Mendelsohn, '"Typhoid Mary" Strikes Again', p. 274.

99. Graham, *The Typhoid Epidemic*, pp. 6–7.

100. 'Social Life at a Standstill in Town of 250,000: Death Bulletin is Posted', *Daily Express*, 20 November 1937, County Borough of Croydon Typhoid Outbreak, Croydon press cuttings.

101. Lords Sitting, Lord Bishop of Winchester, 8 December 1937, Hansard, HC 107.

102. *Evening News*, 16 November 1937, CBC, Typhoid Outbreak 1937–8, Croydon press cuttings, vol. 2.

103. 'Typhoid Fever Outbreak in South Croydon', *Croydon Advertiser*, 6 November 1937, p. 1.

104. Rimington, Typhoid Inquiry, 6 January 1938, p. 745.

105. Holden, Typhoid Inquiry, 28 December 1937, p. 307.

106. Lyons, Typhoid Inquiry, 28 December 1937, p. 307.

107. Holden, Typhoid Inquiry, 28 December 1937, p. 306.

108. S.A. Maycock, Typhoid Inquiry, 6 January 1938, p. 703.

109. Evans, *Death in Hamburg*, pp. 285, 290, 304–5, 380; Snowden, *Naples in the Time of Cholera*, pp. 2, 64, 246–67, 333.

110. Prescott, 'Sending their Sons into Danger', p. 273.

111. 'Social Life at a Standstill', *Daily Express*, 20 November 1937.

112. 'Typhoid Cases outside Croydon: Three Members of Family Ill: Watercress Blamed', *Morning Advertiser*, 20 November 1937, Newspaper cuttings, vol. 1, p. 57; 'Typhoid: Radio Warning', *Daily Herald*, 20 November 1937, Newspaper cuttings, vol. 1, p. 59; 'Typhoid: Radio Watercress SOS', 20 November 1937, Newspaper Cuttings, vol. 1, p. 59; 'Typhoid SOS Stops the Sale of Watercress at Covent Garden', *Evening Standard*, 20 November 1937, Newspaper cuttings, vol. 1, p. 83; 'Typhoid: 13 New Cases, Watercress Scare and Secrecy, Growers Protest, "One of the Safest Crops"', *Star*, 20 November 1937,

Newspaper cuttings, vol. 1, p. 85; 'Typhoid Toll 126: No Clue to Outbreak', *Sunday Chronicle*, 21 November 1937, Newspaper cuttings, vol. 1, p. 87; 'Typhoid Spreads to London. Cress Sellers are Warned', *Sunday Pictorial*, 21 November, Newspaper cuttings, vol. 1, p. 87; 'Typhoid Terror Mystery "Carrier" May Cause New Cases: Secret Probe By Ministry Follows SOS', *Reynolds News*, 21 November 1937, Newspaper cuttings, vol. 1, p. 89; '4 Typhoid Cases in Kensington: 1 Fatal', *Sunday Express*, 21 November 1937, Newspaper cuttings, vol. 1, p. 93; 'More Typhoid Cases, Two London Deaths, "Suspicions" of Watercress" Sales Drop Heavily', *Manchester Guardian*, 22 November 1937, Newspaper cuttings, vol. 1, p. 109.

113. '21 More Cases of Typhoid – 3 Outside Croydon: B.B.C.'s Watercress Warning', *Daily Telegraph*, 20 November 1937, Newspaper cuttings, vol. 1, p. 53.

114. 'Social Life at Standstill in Town of 250,000', *Daily Express*, 20 November 1937, Newspaper cuttings, vol. 1, p. 63.

115. 'Letters to the Editor', *Croydon Advertiser*, letters dated 13 November 1937, p. 10.

116. Ibid., pp. 10–12.

117. 'Letters from Readers', *Croydon Advertiser*, letters dated 20 November 1937, p. 10.

118. 'Tory "Economy" Blamed for Contaminated Well', *Daily Worker*, 20 November 1937; 'Eight Million Londoners Threatened by Epidemic', *Daily Worker*, 22 November 1937; 'Call for Action by Communists in Typhoid Area', *Daily Worker*, 25 November 1937, Newspaper cuttings, vol. 1, pp. 63, 99, 147.

119. 'The Typhoid Tories', *Daily Worker*, 1 March 1938, Croydon press cuttings, vol. 1, p. 2.

120. 'The Typhoid Outbreak Nearing Its End', *Croydon Advertiser*, 10 December 1937, p. 11.

121. CLSL, 'Croydon Typhoid Outbreak 1937' (1995), LS/95/590, S70 (614.4) (small pamphlet briefly describing events and sources available).

122. Murphy, Typhoid Inquiry, 21 December 1937, pp. 94–5.

123. Commons sitting, Major-General Sir Alfred Knox, Wycombe, Conservative, and Sir Patrick Donner, Basingstoke, Conservative, questions to Sir Kingsley Wood, 2 December 1937, Hansard, HC 329.

124. 'They Ate Cress at Croydon Banquet', *Daily Express*; 'They're Not Afraid of Watercress', *Daily Sketch*; 'They've Got the Courage of Their Convictions', *Daily Mirror*; 'Seventeen New Typhoid Cases: Another Death', *News Chronicle*, 25 November 1937, Newspaper cuttings, vol. 1, pp. 157–63.

125. Commons sitting, Craven Ellis, Southampton, Conservative Party, question to Herwald Ramsbotham, Lancaster, 6 December 1937, Hansard, HC 329.

126. 'Strong Protest Against Alarmist Statements', *Croydon Times*, 27 November 1937, Newspaper cuttings, vol. 1, p. 177.

127. 'Typhoid Peril Abating: Shop, Play, Travel as Usual', *Daily Mail*, 24 November 1937, Newspaper cuttings, vol. 1, p. 143.

128. 'Typhoid Shadow Grows Deeper', *Croydon Advertiser*, 26 November 1937, p. 5.

129. 'Strong Protests Against Alarmist Typhoid Statements', *Croydon Times*, 27 November 1937, Newspaper cuttings, vol. 1, p. 177.

130. 'No Reason for Mass Hysteria', *Croydon Advertiser*, 26 November 1937, p. 167.

131. Lord Dawson of Penn and Kaye Le Fleming, 'Letters to the Editor', *The Times*, 22 November 1937, p. 13; 2 December 1937, p. 15; Lord Dawson of Penn, *Interim Report on the Future Provision of Medical and Allied Services, 1920* (London: HMSO, 1920).

132. 'Christmas Shopping is Quite Safe in Croydon – Official', *Croydon Advertiser*, 26 November 1937, Newspaper cuttings, vol. 1, p. 167.

133. Ibid.

134. 'How Medical Detectives Work', *The Star*, 20 November 1937, Newspaper cuttings, vol. 1, p. 167.

135. 'Letters to the Editor', *Croydon Advertiser*, 26 November 1937, p. 23.

136. 'Letters to the Editor', *Croydon Advertiser*, 10 December 1937, p. 15; 17 December 1937, p. 5.

137. Test Action; Finance Committee Meeting, 30 October 1939, Finance Committee Minute Book, 1 November 1938 to 31 October 1939, CAS, CBC.

138. Test Action, p. 77 (snail); pp. 82–3 (lead pipes); pp. 84–5 (tombstone); p. 216 (lack of medicines on a ship); pp. 428–9 (milk).

139. Ministry of Health, *Report on a Public Local Inquiry*, p. 9.

140. Ibid., pp. 10–11.

141. 'Plain Speaking at Special Council Meeting', *Croydon Advertiser*, 4 March 1938, Croydon press cuttings, vol. 2, pp. 147–50.

142. 'Typhoid Demand – "Sack Committee"', *Daily Express*; 'Jeers at Typhoid Meeting', *Daily Herald*; 'Typhoid Committee Offer to Resign', *Daily Mail*; 'Croydon Water Committee May Resign', *Daily Telegraph*; 'Croydon Typhoid Outbreak: Residents' Demands to Council', *The Times*; 'Croydon Committee Offer to Resign', *Daily Sketch*, all articles dated 2 March 1938, Newspaper cuttings, vol. 15, no page numbers.

143. C. Crawford, 'Patients' Rights and the Law of Contract in Eighteenth-Century England', *Social History of Medicine*, 13 (2000), pp. 381–410.

144. See p. 136.

145. 'Typhoid Writs', *Croydon Advertiser*, 8 April 1938, Croydon press cuttings, vol. 2, p. 161.

146. F. Furedi and T. C. Brown, 'Complaining Britain', *Society*, 36 (1999), pp. 72–8, on pp. 73–4. Also see p. 72 for how the British press has presented the culture as being new, especially compared to the US.

147. Ibid., p. 75.

148. See J. Stapleton, *Disease and the Compensation Debate* (Oxford: Oxford University Press, 1986), p. 60 for definitions of test cases and class action.

149. Justice Stable, Test Action, 2 December 1938, p. 458.

150. Furedi and Brown, 'Complaining Britain', p. 76.

151. Ibid., p. 76. They discuss the 1980s and 1990s at this point, though it is not absolutely clear to which period they are referring.

152. Finance Committee Meeting, 16 January 1939, Finance Committee Minute Book, 1 November 1938 to 31 October 1939, CAS, CBC.

153. 'Typhoid Protest', *Daily Mail*, 2 April 1938, Croydon press cuttings, vol. 2, p. 161.

154. 'Sixpence Increase in Rates', *Croydon Advertiser*, 19 March 1938, Croydon press cuttings, vol. 2, p. 159.

155. 'Typhoid Outbreak on the Wane', *Croydon Advertiser*, 3 December 1937, p. 1.

156. Currency converter at http://www.nationalarchives.gov.uk/currency [accessed 1 February 2012], which can convert 1935 sterling to 2005 sterling.

157. Finance Committee Meeting, 30 October 1939, Finance Committee Minute Book, 1 November 1938 to 31 October 1939, CAS, CBC.

158. Wallington, Test Action, 10 November 1938, p. 46; Alfred Read, 11 November 1938, p. 90.

159. 'Typhoid Outbreak on the Wane', *Croydon Advertiser*, 3 December 1937, p. 1.

160. Taylor and Trentmann, 'Liquid Politics'.

161. Murphy and Holden, Typhoid Inquiry, 29 December 1937, p. 396.

162. Finance Committee Meeting, 8 December 1938, Finance Committee Minute Book, 1 November 1938 to 31 October 1939, CAS, CBC.

163. See for example Stapleton, *Disease and the Compensation Debate*; D. J. Ibbetson, *A Historical Introduction to the Law of Obligations* (Oxford: Oxford University Press, 1999).

164. Stapleton, *Disease and the Compensation Debate*, pp. 15, 112.

165. Ibid., pp. 33–4, 38, 50, 61–8.

166. Holden quoted in *Sunday Dispatch*, 28 November 1937, Croydon press cuttings, vol. 2.

167. Honigsbaum, *The Division in British Medicine*, pp. 64–6, 148, 192–4, 206.

168. Dr Forbes questioning Dr Holden, Typhoid Inquiry, 22 December 1937, p. 181, and 28 and 29 December, pp. 328–51.

169. Forbes, Typhoid Inquiry, 29 December 1937, pp. 343, 350.

170. Forbes and Holden, Typhoid Inquiry, 29 December 1937, p. 351.

171. Murphy, Typhoid Inquiry, 29 December 1937, p. 352.

172. '17 New Cases of Typhoid in Croydon Area', *Daily Telegraph*, 25 November 1937, Newspaper cuttings, vol. 1, p. 159.

173. 'Letter to the Editor', *The Times*, 1 December 1937, p. 12.

174. 'Letter to the Editor', *The Times*, 7 February 1938, p. 10.

175. 'Typhoid Relief Fund', *Croydon Advertiser*, 4 March 1938.

176. M. Gorsky, J. Mohan and M. Powell, 'The Financial Health of Voluntary Hospitals in Interwar Britain', *Economic History Review*, 55 (2002), pp. 533–57, on pp. 547, 549–50, 554.

177. Smith and Diack, *Food Poisoning*, p. 123.

178. 20, St Augustine's Avenue (1934 rateable value of £45), Sales Description, Auctioneers Harold Williams, Holliday & Partners, 25 September 1934, CAS, CR(M)79. 40, 42, St Augustine's Avenue (1934 rateable values £64 and £66), Sales Description, Auctioneers Hooker & Rogers in conjunction with Mr Charles Lewin, 31 May 1928, CAS, H+R1986.

179. Rimington's house was no. 49, and Green's was no. 51 and had a rateable value of £52 in 1934. Read lived at 44, Croham Park Avenue, South Croydon. See Croydon Borough Council Valuation List, 1934, South Ward, vol. 14, CAS, St. Augustine's Avenue and Croham Park Avenue, and also for examples of the rateable values other houses in this area. See Stanhope Road for examples of the high valuations.

180. Ministry of Health, *Report on a Public Local Inquiry*, p. 7.

181. 'Typhoid Outbreak', *Croydon Advertiser,* 13 November 1937, p. 1.

182. Smith and Diack, *Food Poisoning*, p. 156.

183. Ibid.

184. Statement for 20 November 1937, Whitgift School Archives (hereafter WSA), SM21/3g.

185. Statement for 27 November 1937, WSA, SM21/3g.

186. Statement for 20 November 1937, WSA, SM21/3g.

187. See p. 132.

188. Stable and Wallington, Test Action, 10 November 1938, p. 47.

189. S. Gates, Monckton and Read, Test Action, 11 November 1938, pp. 87–90.

190. Monckton and Eyre, Test Action, 11 November 1938, p. 101.

191. G. Buchan, Test Action, 16 November 1938, p. 272–5.

192. Stable, Test Action, 16 November 1938, p. 275.

193. Stable, Test Action, 2 December 1938, p. 476.

194. 'Bishop Prays for Victims as Typhoid Toll Mounts', *Sunday Referee*, 21 November 1937, Croydon press cuttings, vol. 1, p. 91.

195. Statement for 18 November 1937, WSA, SM21/3g.
196. 'Football', *Whitgiftian*, 55 (December 1937), pp. 121–31.
197. F. H. G. Percy, *Whitgift School: A History* (Croydon: The Whitgift Foundation, 1991, 1st edn 1976), p. 184.
198. A. Warwick, *Masters of Theory: Cambridge and the Rise of Mathematical Physics* (Chicago, IL, and London: The University of Chicago Press, 2003), pp. 178, 213.
199. Prescott, 'Sending their Sons into Danger', p. 308.
200. See p. 162.
201. Political and Economic Planning, *Report on the British Health Services*, p. 156.
202. Ministry of Health, *Report on a Public Local Inquiry*, p. 13.
203. Dawson and Sartory, 'Microbiological Safety of Water', p. 75.
204. Mendelsohn, '"Typhoid Mary" Strikes Again', p. 273.
205. A. Mendelsohn, 'From Eradication to Equilibrium: How Epidemics became Complex after World War I', in C. Lawrence and G. Weisz, *Greater than the Parts: Holism in Biomedicine, 1920–1950* (Oxford: Oxford University Press, 1998), pp. 303–31.

Conclusion

1.   Mrs H. to Malton Typhoid Relief Committee, 2 May 1933, NYRO TRF.
2.   P. Mandler, 'Against "Englishness": English Culture and the Limits to Rural Nostalgia, 1850–1940', *Transactions of the Royal Historical Society*, 6th series, 7 (1997), pp. 155–75, on pp. 155–6, 175.
3.   Anne Hardy refers directly to Wiener in her article on diphtheria practices, in which she frames delay in use of intubation with English 'conservatism'. See Hardy, 'Tracheotomy versus Intubation', p. 558. Terrie Romano frames the apathy towards experimental science in the late nineteenth century at Oxford University as due to the cultured, classically educated physicians. See Romano, *Making Medicine Scientific*, pp. 142–3.
4.   J. H. Warner, 'Ideals of Science and their Discontents in Late Nineteenth-Century American Medicine', *Isis*, 82 (1991), pp. 454–78, on p. 454. See also C. E. Rosenberg, 'The Therapeutic Revolution: Medicine, Meaning, and Social Change in Nineteenth-Century America', in C. E. Rosenberg, *Explaining Epidemics and Other Studies in the History of Medicine* (Cambridge: Cambridge University Press, 1992), pp. 9–31, on p. 31, which discusses the 'easy romanticization of [a] community lost' due to the arrival of laboratory medicine; Sturdy, 'Looking for Trouble', p. 747.
5.   Data on Bart's physicians extracted from Brown *Munk's Roll* and Trail, *Munk's Roll*. Comparing only those physicians who became FRCPs before 1889, in line with Peterson's study. FRCP data from Peterson, *Medical Profession*, pp. 49–50.
6.   M. Collins, 'The Fall of the English Gentleman: The National Character in Decline, *c.* 1918–1970', *Historical Research*, 75 (2002), pp. 90–111.
7.   Ibid., p. 102.
8.   Sturdy, 'Looking for Trouble', esp. pp. 740–7.
9.   Moore, *St Bartholomew's Hospital, Volume II*, pp. 723–55. The pathologists received about five pages of this chapter, pp. 750–5.
10. Ibid., p. 755.
11. Bodley Scott, 'Medicine in the Twentieth Century'; Medvei and Thornton, *The Royal Hospital of Saint Bartholomew, 1123–1973*. See Waddington, *Medical Education*, pp. 115–45 for a historian's perspective, as discussed in Chapter 1, p. 15.

12. Worboys, *Spreading Germs*; M. Worboys, 'Was there a Bacteriological Revolution in Late Nineteenth-Century Medicine?', *Studies in the History and Philosophy of Biological and Biomedical Sciences*, 38 (2007), pp. 20–42; Tomes, 'The Private Side'; Tomes, *Gospel of Germs*.
13. Weindling, 'From Medical Research to Clinical Practice', p. 78.
14. Stark, 'Bacteriology in the Service of Sanitation', pp. 343–61.
15. For example, E. Dyck and C. Fletcher (eds), *Locating Health: Historical and Anthropological Investigations of Place and Health* (London: Pickering & Chatto, 2011).
16. Rosenberg, *The Cholera Years*.
17. Howell, *Technology in the Hospital*, p. 195.

WORKS CITED

Archives and Special Collections

Addenbrooke's Hospital Archives, AHA, AHM 31, 33, 35, 39, 40, 42, 43, 45.

Addenbrooke's Hospital Archives, AHA, AHPR 1/1, 025/8–9.

Cambridgeshire Collection, Milton Road Library, Cambridge, 'Report of the Sanitary Condition of the Cambridge Improvement Act District', 1880–8, 1889–1900, 1902–20.

Cambridgeshire Collection, Milton Road Library, Cambridge, 'Report on the Sanitary Condition of the Borough of Cambridge, 1907'.

County Record Office, Cambridge, Meetings of the Public Health Committee, 1899, 1901, 1902.

County Record Office, Cambridge, Public Health Committee Minute Book, 1899–1904.

Croydon Local Studies Library and Archives Service, Croydon Borough Council Valuation List, 1934.

Croydon Local Studies Library and Archives Service, Finance Committee Meeting.

Croydon Local Studies Library and Archives Service, Finance Committee Minute Book.

Croydon Local Studies Library and Archives Service, Inquiry into the Outbreak of Typhoid Fever in October and November 1937, Minutes of Proceedings.

Croydon Local Studies Library and Archives Service, Sales Description, 31 May 1928; 25 September 1934.

Croydon Local Studies Library and Archives Service, Test Action, A. and P. R. Read v. Croydon Corporation, November 1938.

Croydon Local Studies Library and Archives Service, 'Newspaper Cuttings Relating to the Typhoid Outbreak in Croydon', vols 1, 2, 15.

Gwynedd Archives, Caernarfon Record Office, School Log Books.

International Labour Organization Archive, Advisory Committee on Anthrax.

International Labour Organization Archive, International Anthrax Commission, *Memorandum*.

International Labour Organization Archive, Letters and Draft Letters.

League of Nations Archives, United Nations, Geneva, Industrial Hygiene, Enquiry and Correspondence.

Marlborough College Archives, Bursar Thomas 1889–90.

Marlborough College Archives, Medical Box 331.

Marlborough College Archives, Minutes of General Meetings, 1845–93.

Marlborough College Archives, Unsorted Letters, Bursar Tomkinson, 1858.

North Yorkshire County Record Office, Typhoid Relief Fund, ZPB

Royal College of Physicians, London, S. Gee, MS 34, 35, 36.

St Bartholomew's Hospital Archives, London, Medical Council Minutes.

St Bartholomew's Hospital Archives, London, Medical Registers.

St Bartholomew's Hospital Archives, London, Meetings.

St Bartholomew's Hospital Archives, London, Minutes.

St Bartholomew's Hospital Archives, London, Statistical Tables.

St Bartholomew's Hospital Archives, London, Treasurer's Reports.

University of Bradford Special Collections, A15, Newspaper cuttings.

University of Wales, Bangor, Department of Archives and Manuscripts, Florence Nightingale letters, MSS 37616, 37617, 37619.

Wellcome Library Manuscripts and Archives, Letters, GP/31.

West Yorkshire Archive Service, Bradford, Bradford Medico-Chirurgical Society Minute Book, 1874–84.

West Yorkshire Archive Service, Bradford, National Union of Woolsorters' Minutes, 1898–1971.

West Yorkshire Archive Service, 'Notebook Belonging to Dr. Bell'.

Whitgift School Archives, Croydon, Statements.

Zoological Society of London Archives, Zoological Society Minutes of Council, 1926–9.

Periodicals and Newspapers

Bradford Daily Telegraph

Bradford Observer

British Medical Journal

Canadian Public Health Journal

Carnarvon and Denbigh Herald

Croydon Advertiser

Eccles and Patricroft Journal

Guy's Hospital Reports

Kent Messenger

Lancet

Leather

Leather World

Malburian

Malton Messenger

Occupational Health and Environmental Medicine

Proceedings of the Royal Society of Medicine

Proceedings of the Zoological Society of London

Rotherham Advertiser

South Eastern Gazette

South London Press

St Bartholomew's Hospital Journal

The Times

Transactions of the Epidemiological Society of London

Whitgiftian

Yorkshire Daily Observer

Yorkshire Herald

Yorkshireman

Government and Other Reports

Borough of Bradford

Hansard

Home Office

International Labour Office

Local Government Board

London County Council

Ministry of Health

North Riding of Yorkshire County Council

Urban District Council of Malton

World Health Organization

Published Primary Sources

Bernard, C., 'An Introduction to the Study of Experimental Medicine' (1865), in C. Bernard, *Experimental Medicine* with a new introduction by S. Wolf, trans. H. C. Greene (New Brunswick, NJ, and London, Transaction Publishers, 1999), pp. 1–226.

Bradford Medico-Chirurgical Society, *Report of the Commission on Woolsorter's Disease* (Bradford, 1882).

Crookshank, E. M., *A Text-Book of Bacteriology including the Etiology and Prevention of Infective Diseases*, 4th edn (London: H. K. Lewis, 1896).

Dawson, B., Lord, *Interim Report on the Future Provision of Medical and Allied Services, 1920* (London: HMSO, 1920).

Ferris, P., *The Doctors* (London: Victor Gollancz Ltd, 1965).

Ford, W. W., *Bacteriology* (New York and London: Paul B. Hoeber Inc., 1939).

Gee, S., *Medical Lectures and Aphorisms*, with recollections by J. W. Legg (London: Henry Frowde, Oxford University Press and Hodder and Stoughton, 1902, 1915).

Horder, T. J., *Clinical Pathology in Practice* (London: Henry Frowde, Hodder and Stoughton and Oxford University Press, 1910).

—, *Medical Notes* (London: Henry Frowde and Hodder and Stoughton, 1921).

—, *Health and a Day: Addresses by Lord Horder* (London: J. M. Dent & Sons Ltd, 1937).

Horder, T. J. and A. E. Gow, *The Essentials of Medical Diagnosis: A Manual for Students and Practitioners*, rev. with the assistance of R. Bodley Scott (London: Toronto, Melbourne and Sydney: Cassell and Company Ltd, 1952).

International Labour Office, 'The Prevention of Anthrax: Draft Regulations for Protection Against Infection by Anthrax in the Hides and Skins Industry', in International Labour Office, *The International Labour Code, 1939* (Montreal: International Labour Office, 1941), pp. 590–9.

Legg, J. W., 'Recollections of Samuel Gee, Physician to Saint Bartholomew's Hospital brought together by J. W. Legg' (1911), in S. Gee, *Medical Lectures and Aphorisms*, with recollections by J. W. Legg (London: Henry Frowde, Oxford University Press and Hodder and Stoughton, 1915), pp. 353–92.

The Newspaper Press Directory and Advertisers' Guide, Containing Full Particulars of Every Newspaper, Magazine, Review, and Periodical Published in the United Kingdom and the British Isles with the Newspaper Map of the United Kingdom, the Principal Continental and American Papers, and a Directory of the Class Papers and Periodicals, Twenty-Fifth Annual Issue (London: C. Mitchell and Co. Advertising Contractors, 1880).

Political and Economic Planning, *Report on the British Health Services: A Survey of the Existing Health Services in Great Britain with Proposals for Future Development, December 1937* (London: PEP, 1937).

Ponder, C., *A Report to the Worshipful Company of Leathersellers on the Incidence of Anthrax amongst those Engaged in the Hide, Skin and Leather Industries, with an Inquiry into Certain Measures Aiming at its Prevention* (London: Worshipful Company of Leathersellers, 1911).

Price, E. O., 'The Bangor Typhoid Epidemic of 1882', MD thesis, 1891, reprinted in *Caernarvonshire Historical Society Transactions*, 26 (1965), pp. 157–68.

Scott, H. H., 'Tuberculosis in Captive Wild Animals as Compared and Contrasted with the Disease in Man', *Proceedings of the Royal Society of Medicine*, 20 (1927), pp. 197–204.

—, *Tuberculosis in Man and Lower Animals: A Study in Comparative Pathology*, Medical Research Council, Special Report Series, No. 149 (1930).

Shaw, W. V., 'Report on an Outbreak of Enteric Fever in the County Borough of Bournemouth and in the Boroughs of Poole and Christchurch', *Reports on Public Health and Medical Subjects*, 81 (London: HMSO, 1937).

Tibbits, E. T., *Medical Fashions in the Nineteenth Century, including a Sketch of Bacteriomania and the Battle of the Bacilli* (London: H. K. Lewis, 1884).

Willis, W. A., *The Workmen's Compensation Act, 1906* (London: Butterworth and Co. and Shaw and Sons, 1913).

Secondary Sources

Abel-Smith, B., *The Hospitals 1800–1948: A Study in Social Administration in England and Wales* (London: Heinemann, 1964).

Ackerknecht, E. H., 'A Plea for a "Behaviourist" Approach in Writing the History of Medicine', *Journal of the History of Medicine and Allied Sciences*, 22 (1967), pp. 211–14.

Alborn, T., 'Insurance against Germ Theory: Commerce and Conservatism in Late-Victorian Medicine', *Bulletin of the History of Medicine*, 75 (2001), pp. 406–45.

Alvin, C., 'Medical Treatment and Care in Nineteenth-Century Bradford: An Examination of Voluntary, Statutory, and Private Medical Provision in a Nineteenth-Century Urban Industrial Community' (PhD dissertation, University of Bradford, 1998).

Barnes, E., 'Fashioning a Natural Self: Guides to Self-Fashioning in Victorian England' (PhD thesis, University of Cambridge, 1996).

Bartrip, P., *Workmen's Compensation in Twentieth Century Britain: Law, History and Social Policy* (Aldershot: Avebury, 1987).

—, *The Home Office and the Dangerous Trades: Regulating Occupational Disease in Victorian and Edwardian Britain* (Amsterdam: Rodopi, 2002).

BBC News, 'Anthrax Outbreak Hits Bangladesh Leather and Meat Sectors', 13 October 2010, at www.bbc.co.uk/news/business-11451570 [accessed 30 October 2012].

Bligh, M., *Dr. Eurich of Bradford* (London: James Clarke & Co. Ltd, 1960).

Bodley Scott, R., 'Medicine in the Twentieth Century', in V. C. Medvei and J. L. Thornton (eds), *Royal Hospital of Saint Bartholomew 1123–1973* (London: Saint Bartholomew's Hospital, 1974), pp. 185–204.

Bourne, G., *We met at Bart's: The Autobiography of a Physician* (London: Friedrich Muller Limited, 1963).

Briggs, A., 'Cholera and Society in the Nineteenth Century', in A. Briggs, *The Collected Essays of Asa Briggs, Volume II: Images, Problems, Standpoints, Forecasts* (Brighton: The Harvester Press, 1985), pp. 153–76.

Brown, G. H. (ed.), *Munk's Roll, Volume IV: Lives of the Fellows of the Royal College of Physicians of London, 1826–1925* (London: Royal College of Physicians, 1955).

Bud, R., *Penicillin: Triumph and Tragedy* (Oxford: Oxford University Press, 2007).

Bulloch, W., *The History of Bacteriology* (London: Oxford University Press, 1938).

Canguilhem, G., *The Normal and the Pathological* (New York: Zone Books, 1991).

Cantor, D., 'The MRC's Support for Experimental Radiology during the Inter-War Years', in J. Austoker and L. Bryder (eds), *Historical Perspectives on the Role of the MRC* (Oxford: Oxford University Press, 1989), pp. 181–204.

Cantor, N. F., *In the Wake of the Plague: Black Death and the World it Made* (New York: Harper Collins, 2002).

Carter, J. T., 'Anthrax in Kidderminster, 1900–1914' (PhD dissertation, University of Birmingham, 2005).

Carter, K. C., 'The Koch–Pasteur Dispute on Establishing the Cause of Anthrax', *Bulletin of the History of Medicine*, 62 (1988), pp. 42–57.

Carter, T. and J. Melling, 'Trade, Spores, and the Culture of Disease: Attempts to Regulate Anthrax in Britain in Its International Trade, 1875–1930', in C. Sellers and J. Melling, *Dangerous Trade: Histories of Industrial Hazard Across a Globalizing World* (Philadelphia, PA: Temple University Press, 2012) pp. 60–72.

Clark, J. F. M., *Bugs and the Victorians* (New Haven, CT, and London: Yale University Press, 2009).

Coleman, W., *Yellow Fever in the North: The Methods of Early Epidemiology* (Madison, WI: The University of Wisconsin Press, 1987).

Collins, M., 'The Fall of the English Gentleman: The National Character in Decline, c. 1918–1970', *Historical Research*, 75 (2002), pp. 90–111.

Condrau, F. and M. Worboys, 'Second Opinions: Epidemics and Infections in Nineteenth-Century Britain', *Social History of Medicine*, 20 (2007), pp. 147–58.

Cranefield, P. F., 'The Organic Physics of 1847 and the Biophysics of Today', *Journal of the History of Medicine and Allied Sciences*, 12 (1957), pp. 407–23.

Crenner, C., 'Professional Measurement: Quantifying Health and Disease in American Medical Practice, 1880–1920' (PhD dissertation, Harvard University, 1993).

—, 'Diagnosis and Authority in the Early Twentieth-Century Medical Practice of Richard C. Cabot', *Bulletin of the History of Medicine*, 76 (2002), pp. 30–55.

—, *Private Practice: In the Early Twentieth-Century Medical Office of Dr Richard Cabot* (Baltimore, MD: The Johns Hopkins University Press, 2005).

Cunningham, A. and P. Williams, 'Introduction', in A. Cunningham and P. Williams (eds), *The Laboratory Revolution in Medicine* (Cambridge: Cambridge University Press, 2002), pp. 1–13.

Dale, H., 'Scientific Method in Medical Research. An Address Given on October 10, 1950, in Opening a Course of Lectures on "The Scientific Basis of Medicine" Arranged by the British Postgraduate Medical Federation', *British Medical Journal* (1950), pp. 1185–90.

Daston, L. and H. O. Sibum, 'Introduction: Scientific Personae and their Histories', *Science in Context*, 16 (2003), pp. 1–8.

Davis, G. L., *'The Cruel Madness of Love': Sex, Syphilis and Psychiatry in Scotland, 1880–1930* (Amsterdam and New York: Rodopi, 2008).

Delaporte, F., *Disease and Civilisation: The Cholera in Paris, 1832*, trans. A. Goldhammer (Cambridge, MA, and London: The MIT Press, 1986).

Digby, A., *The Evolution of British General Practice 1850–1948* (Oxford: Oxford University Press, 1999).

Digby, A. and N. Bosanquet, 'Doctors and Patients in an Era of National Health Insurance and Private Practice, 1913–1938', *Economic History Review*, new series, 41 (1988), pp. 74–94.

Edgerton, D., *Science, Technology and the British Industrial 'Decline' 1870–1970* (Cambridge: Cambridge University Press, 1996).

—, 'From Innovation to Use: Ten Eclectic Theses on the Historiography of Technology', *History and Technology*, 16 (1999), pp. 111–36.

Epstein, S., *Impure Science: AIDS, Activism, and the Politics of Knowledge* (Berkeley, CA: University of California Press, 1996).

Ernst, W., 'The Normal and the Abnormal: Reflections on Norms and Normativity', in W. Ernst, *Histories of the Normal and the Abnormal: Social and Cultural Histories of Norms and Normativity* (Abingdon and New York: Routledge, 2006), pp. 1–25.

Evans, R. J., *Death in Hamburg: Society and Politics in the Cholera Years, 1830–1910* (Oxford: Clarendon Press, 1987).

Figlio, K., 'What is an Accident?', in P. Weindling (ed.), *The Social History of Occupational Health* (London: Croom Helm, 1985), pp. 180–206.

Firth, G., *A History of Bradford* (Chichester: Phillimore, 1997).

Foster, W. D., *A History of Medical Bacteriology and Immunology* (London, William Heinemann Medical Books Ltd, 1970).

Frank Jr, R. G., 'The Telltale Heart: Physiological Instruments, Graphic Methods, and Clinical Hopes, 1854–1914', in W. Coleman and F. L. Holmes (eds), *The Investigative Enterprise: Experimental Physiology in Nineteenth-Century Medicine* (Berkeley and Los Angeles, CA, and London, University of California Press, 1988), pp. 211–90.

Fressoz, J. B., 'Beck Back in the 19th Century: Towards a Genealogy of Risk Society', *History and Technology: An International Journal*, 23 (2007), pp. 333–50.

Furedi, F. and T. C. Brown, 'Complaining Britain', *Society*, 36 (1999), pp. 72–9.

Geison, G. L., *Michael Foster and the Cambridge School of Physiology: The Scientific Enterprise in Late Victorian Society* (Princeton, NJ: Princeton University Press, 1978).

—, 'Divided We Stand: Physiologists and Clinicians in the American Context', in M. J. Vogel and C. E. Rosenberg (eds), *The Therapeutic Revolution: Essays in the Social History of American Medicine* (Philadelphia, PA: University of Pennsylvania Press, 1979), pp. 67–90.

Gent, J., *Croydon Past* (Chichester: Phillimore and Co. Ltd, 2002).

Goffman, E., *The Presentation of Self in Everyday Life* (London: Allen Lane The Penguin Press, 1969).

Gorski, R., 'Health and Safety aboard British Merchant Ships: The Case of First Aid Instruction, 1881–1908', in R. Gorski (ed.), *Maritime Labour: Contributions to the History of Work at Sea, 1500–2000* (Amsterdam: Aksant, 2007), pp. 119–40.

Gorsky, M., J. Mohan and M. Powell, 'The Financial Health of Voluntary Hospitals in Interwar Britain', *Economic History Review*, 55 (2002), pp. 533–57.

Gradmann, C., *Laboratory Disease: Robert Koch's Medical Bacteriology* (Baltimore, MD: The Johns Hopkins University Press, 2009).

Graham, M., *The Typhoid Epidemic in Bournemouth, Poole and Christchurch 1936* (Christchurch: Bournemouth Local Studies Publications, 1997).

Gregory, B. S., '*Is* Small Beautiful? Microhistory and the History of Everyday Life', *History and Theory: Studies in the Philosophy of History*, 38 (1999), pp. 100–10.

Guillemin, J., *Anthrax: The Investigation of a Deadly Outbreak* (Berkeley, CA, and London: University of California Press, 1999).

Hamlin, C., *A Science of Impurity: Water Analysis in Nineteenth-Century Britain* (Berkeley, CA: University of California Press, 1990).

Hammerborg, M., 'The Laboratory and the Clinic Divide Revisited: The Introduction of Laboratory Medicine at the Bergen Hospital, Norway', *Social History of Medicine*, 24 (2011), pp. 758–75.

Hammonds, E. M., *Childhood's Deadly Scourge: The Campaign to Control Diphtheria in New York City, 1880–1930* (Baltimore, MD, and London: The Johns Hopkins University Press, 1999).

Hardy, A., 'Tracheotomy versus Intubation: Surgical Intervention in Diphtheria in Europe and the United States, 1825–1930', *Bulletin of the History of Medicine*, 66 (1992), pp. 536–59.

—, 'On the Cusp: Epidemiology and Bacteriology at the Local Government Board, 1890–1905', *Medical History*, 42 (1998), pp. 328–46.

—, 'Food, Hygiene, and the Laboratory: A Short History of Food Poisoning in Britain, *circa* 1850–1950', *Social History of Medicine*, 12 (1999), pp. 293–311.

—, '"Straight Back to Barbarism": Antityphoid Inoculation and the Great War, 1914', *Bulletin of the History of Medicine*, 74 (2000), pp. 265–90.

—, 'Methods of Outbreak Investigation in the "Era of Bacteriology" 1880–1920', *Sozial Präventivmed*, 46 (2001), pp. 355–60.

—, 'Exorcising Molly Malone: Typhoid and Shellfish Consumption in Urban Britain 1860–1960', *History Workshop Journal*, 55 (2003), pp. 73–90.

—, 'Scientific Strategy and Ad Hoc Response: The Problem of Typhoid in America and England, *c.* 1910–50', *Journal of the History of Medicine and Allied Sciences* (April 2012), pp. 1–35.

Holmes, C., *Spores, Plagues and History: The Story of Anthrax* (Dallas, TX, Durban House Pub., 2003).

Honigsbaum, F., *The Division in British Medicine: A History of the Separation of General Practice from Hospital Care 1911–1968* (New York: St Martin's Press, 1979).

Hooker, C. and A. Bashford, 'Diphtheria and Australian Public Health: Bacteriology and its Complex Applications, *c.* 1890–1930', *Medical History*, 46 (2002), pp. 41–64.

Hope Simpson, J. B., *Rugby Since Arnold: A History of Rugby School from 1842* (London and New York: Macmillan and St. Martin's Press, 1967).

Horder, M., *The Little Genius: A Memoir of the First Lord Horder* (London: Gerald Duckworth & Co Ltd, 1966).

Howell, J. D., *Technology in the Hospital: Transforming Patient Care in the Early Twentieth Century* (Baltimore, MD, and London: The Johns Hopkins University Press, 1995).

Hoy, S., *Chasing Dirt: The American Pursuit of Cleanliness* (New York and Oxford: Oxford University Press, 1995).

Hull, A. J., 'Teamwork, Clinical Research, and the Development of Scientific Medicine in Interwar Britain: The Glasgow School Revisited', *Bulletin of the History of Medicine*, 81 (2007), pp. 569–93.

Humphries, M., 'Typhoid and its Carriers', in K. F. Kiple (ed.), *Plague, Pox and Pestilence: Disease in History* (London: Weidenfeld and Nicolson, 1997), pp. 14–19.

Hunter, D., *The Diseases of Occupations* (London: English Universities Press, 1955).

Ibbetson, D. J., *A Historical Introduction to the Law of Obligations* (Oxford: Oxford University Press, 1999).

Irwin, A., *Citizen Science: A Study of People, Expertise and Sustainable Development* (London: Routledge, 1995).

Ittman, K., *Work, Gender and Family in Victorian England* (New York: New York University Press, 1995).

Jacyna, L. S., 'The Laboratory and the Clinic: The Impact of Pathology on Surgical Diagnosis in the Glasgow Western Infirmary, 1875–1910', *Bulletin of the History of Medicine*, 62 (1988), pp. 384–406.

James, D., D. 'William Byles and the *Bradford Observer*', in D. G. Wright and J. A. Jowitt (eds), *Victorian Bradford: Essays in Honour of Jack Reynolds* (Bradford: City of Bradford Metropolitan Council, 1982), pp. 115–36.

—, *Bradford* (Halifax: Ryburn Publishing, 1990).

Jones, S., *Death in a Small Package: A Short History of Anthrax* (Baltimore, MD: The Johns Hopkins University Press, 2010).

Jones, S. and P. Teigen, 'Anthrax in Transit: Practical Experience and Intellectual Exchange', *Isis*, 99 (2008), pp. 455–85.

Keele, K. D., *The Evolution of Clinical Methods in Medicine being the FitzPatrick Lectures Delivered at the Royal College Physicians in 1960–61* (London: Pitman Medical Publishing Co. Ltd, 1963).

Laborde, E. D., *Harrow School, Yesterday and Today* (London: Winchester Publications Limited, 1948).

Latour, B., *The Pasteurization of France*, trans. A. Sheridan and J. Law (Cambridge, MA, and London: Harvard University Press, 1988).

Lawrence, C., 'Incommunicable Knowledge: Science, Technology and the Clinical Art in Britain, 1850–1914', *Journal of Contemporary History*, 20 (1985), pp. 503–20.

—, 'Still Incommunicable: Clinical Holists and Medical Knowledge in Interwar Britain', in C. Lawrence and G. Weisz (eds), *Greater than the Parts: Holism in Biomedicine, 1920–1950* (New York and Oxford: Oxford University Press, 1998), pp. 94–111.

—, 'A Tale of Two Sciences: Bedside and Bench in Twentieth-Century Britain', *Medical History*, 43 (1999), pp. 421–49.

—, 'Edward Jenner's Jockey Boots and the Great Tradition in English Medicine 1918–1939', in C. Lawrence and A. K. Mayer (eds), *Regenerating England: Science, Medicine and Culture in Inter-War Britain* (Amsterdam: Rodopi, 2000), pp. 45–65.

—, *Rockefeller Money, The Laboratory, and Medicine in Edinburgh, 1919–1930: New Science in an Old Country* (Rochester, NY, and Woodbridge: University of Rochester and Boydell and Brewer, 2005).

Laybourn, K. and D. James, 'Introduction', in K. Laybourn and D. James (eds), *The Rising Sun of Socialism: The Independent Labour Party in the Textile District of the West Riding of Yorkshire between 1890 and 1914* (West Yorkshire: West Yorkshire Archive Service, 1991).

Leavitt, J. W., *Typhoid Mary: Captive to the Public's Health* (Boston, MA: Beacon Press, 1996).

LeBaron, C. W. and D. N. Taylor, 'Typhoid Fever', in K. F. Kiple (ed.), *The Cambridge World History of Human Disease* (Cambridge: Cambridge University Press, 1999), pp. 1071–7.

Lindqvist, S., 'Change in the Technological Landscape: The Temporal Dimension in the Growth and Decline of Large Technological Systems', in O. Granstrand (ed.), *Economics of Technology* (Amsterdam: Elsevier Science B.V., 1994), pp. 271–88.

McDonald, L. 'Mythologizing and De-Mythologizing', in S. Nelson and A. M. Rafferty (eds), *Notes on Nightingale* (Ithaca, NY: Cornell University Press, 2010), pp. 91–114.

McGrew, R. E., *Russia and the Cholera 1823–1832* (Madison and Milwaukee, WI: The University of Wisconsin Press, 1965).

McKay, R., 'Imagining "Patient Zero": Sexuality, Blame, and the Origins of the North American AIDS Epidemic' (PhD thesis, University of Oxford, 2010).

Mandler, P., 'Against "Englishness": English Culture and the Limits to Rural Nostalgia, 1850–1940', *Transactions of the Royal Historical Society*, 6th series, 7 (1997), pp. 155–75.

—, *The Fall and Rise of the English Stately Home* (New Haven, CT, and London: Yale University Press, 1997).

Matthews David, A., 'Made to Measure? Tailoring and the "Normal" Body in Nineteenth-Century France', in W. Ernst, *Histories of the Normal and the Abnormal: Social and Cultural Histories of Norms and Normativity* (Abingdon and New York: Routledge, 2006), pp. 142–64.

Maulitz, R. C., '"Physician Versus Bacteriologist": The Ideology of Science in Clinical Medicine', in M. J. Vogel and C. E. Rosenberg (eds), *The Therapeutic Revolution: Essays in the Social History of American Medicine* (Philadelphia, PA: University of Pennsylvania Press, 1979), pp. 91–107.

Medick, H., '"Missionaries in the Rowboat"? Ethnological Ways of Knowing as a Challenge to Social History', in A. Lüdtke (ed.), *The History of Everyday Life: Reconstructing Historical Experiences and Ways of Life*, trans. W. Templar (Princeton, NJ, Princeton University Press, 1995), pp. 41–71.

Melling, J., 'Beyond a Shadow of a Doubt? Experts, Lay Knowledge, and the Role of Radiography in the Diagnosis of Silicosis in Britain, c. 1919–1945', *Bulletin of the History of Medicine*, 84 (2010), pp. 424–66.

Melling, J. and C. Sellers, 'Objective Collectives? Transnationalism and "Invisible Colleges" in Occupational and Environmental Health from Collis to Selikoff', in C. Sellers and J. Melling, *Dangerous Trade: Histories of Industrial Hazard Across a Globalizing World* (Philadelphia, PA: Temple University Press, 2012), pp. 113–25.

Mendelsohn, J. A., '"Typhoid Mary" Strikes Again: The Social and the Scientific in the Making of Modern Public Health', *Isis*, 86 (1995), pp. 268–77.

—, 'Cultures of Bacteriology: Formation and Transformation of a Science in France and Germany, 1870–1914' (PhD thesis, Princeton University, 1996).

—, 'From Eradication to Equilibrium: How Epidemics Became Complex after World War I', in C. Lawrence and G. Weisz (eds), *Greater than the Parts: Holism in Biomedicine, 1920–1950* (New York and Oxford: Oxford University Press, 1998), pp. 303–31.

Metcalfe, N., 'The History of Woolsorters' Disease: A Yorkshire Beginning with an International Future?', *Occupational Medicine*, 54 (2004), pp. 489–93.

Moore, N., *The History of St. Bartholomew's Hospital, Volume II* (London: C. Arthur Pearson Limited, 1918).

Mortimer, I. and J. Melling, '"The Contest between Commerce and Trade, on the One Side, and Human Life on the Other": British Government Policies for the Regulation of Anthrax Infection and the Wool Textiles Industries, 1880–1939', *Textile History*, 31:2 (2000), pp. 222–36.

Moses, J., 'Contesting Risk: Specialist Knowledge and Workplace Accidents in Britain, Germany, and Italy, 1870–1920', in K. Brückweh, D. Schumann, R. F. Wetzell and B. Ziemann, *Engineering Society: the Role of the Human and Social Sciences in Modern Societies, 1880–1980* (Basingstoke: Palgrave, 2012), pp. 59–78.

Nicoll, A. and R. Maynard, 'One Hundred Years of Anthrax', *Occupational and Environmental Medicine*, 61 (2004), p. 95.

Palladino, P., 'On Writing the Histor(ies) of Modern Medicine', *Rethinking History*, 3 (1999), pp. 271–88.

Pasveer, B., 'Depiction in Medicine as a Two-Way Affair: X-Ray Pictures and Pulmonary Tuberculosis in the Early Twentieth Century', in I. Löwy (ed.), *Medicine and Change: Historical and Sociological Studies of Medical Innovation. Proceedings of the Symposium INSERM held in Paris, 21–23 April, 1992* (London and Paris: John Libbey Eurotext and INSERM, 1993).

Pelling, M., 'Contagion/Germ Theory/Specificity', in W. F. Bynum and R. Porter (eds), *Companion Encyclopaedia of the History of Medicine, Volume 1* (London and New York: Routledge, 1993), pp. 309–34.

Pemberton, N. and M. Worboys, *Mad Dogs and Englishmen: Rabies in Britain, 1830–2000* (Basingstoke: Palgrave Macmillan, 2007).

—, *Rabies in Britain: Dogs, Disease and Culture, 1830–2000* (Basingstoke: Palgrave Macmillan, 2012).

Percy, F. H. G., *Whitgift School: A History* (Croydon: The Whitgift Foundation, 1991, 1st edn 1976).

Peterson, M. J., *The Medical Profession in Mid-Victorian London* (Berkeley and Los Angeles, CA, and London: University of California Press, 1978).

Porter, R., *Bodies Politic: Disease, Death and Doctors in Britain, 1650–1900* (Ithaca, NY: Cornell University Press, 2001).

Prescott, H., 'Sending Their Sons into Danger: Cornell University and the Ithaca Typhoid Epidemic of 1903', *New York History*, 78 (1997), pp. 273–308.

Richardson, N., *Typhoid in Uppingham* (London: Pickering & Chatto, 2008).

Risse, G. B. and J. H. Warner, 'Reconstructing Clinical Activities: Patient Records in Medical History', *Social History of Medicine*, 5 (1992), pp. 183–205.

Robins, J., *The Miasma: Epidemic and Panic in Nineteenth Century Ireland* (Dublin: Institute of Public Administration, 1995).

Rogers, N., 'Germs with Legs: Flies, Disease, and the New Public Health', *Bulletin of the History of Medicine*, 63 (1989), pp. 599–617.

Romano, T. M., *Making Medicine Scientific: John Burdon Sanderson and the Culture of Victorian Science* (Baltimore, MD, and London: The Johns Hopkins University Press, 2002).

Rook, A., M. Carlton and W. G. Cannon, *The History of Addenbrooke's Hospital, Cambridge* (Cambridge: Cambridge University Press, 1991).

Rosenberg, C. E., *The Cholera Years: The United States in 1832, 1849, and 1866* (Chicago, IL, and London: University of Chicago Press, 1962).

—, 'Florence Nightingale on Contagion: The Hospital as Moral Universe', in C. E. Rosenberg (ed.), *Explaining Epidemics and Other Studies in the History of Medicine* (New York: Cambridge University Press, 1992), pp. 116–36.

—, 'The Therapeutic Revolution: Medicine, Meaning, and Social Change in Nineteenth-Century America', in C. E. Rosenberg, *Explaining Epidemics and Other Studies in the History of Medicine* (Cambridge, Cambridge University Press, 1992), pp. 9–31.

Sinding, C., 'The Power of Norms: Georges Canguilhem, Michel Foucault, and the History of Medicine', in F. Huisman and J. H. Warner (eds), *Locating Medical History: The Stories and their Meanings* (Baltimore, MD, and London: The Johns Hopkins University Press, 2004), pp. 262–84.

—, 'Flexible Norms? From Patients' Values to Physicians' Standards', in W. Ernst, *Histories of the Normal and the Abnormal: Social and Cultural Histories of Norms and Normativity* (Abingdon and New York: Routledge, 2006), pp. 225–44.

Smith, D. F. and H. L. Diack with T. H. Pennington and E. M. Russell, *Food Poisoning, Policy and Politics: Corned Beef and Typhoid in Britain in the 1960s* (Woodbridge: The Boydell Press, 2005).

Snowden, F. M., *Naples in the Time of Cholera, 1884–1911* (Cambridge: Cambridge University Press, 1995).

Stark, J., 'Industrial Illness in Cultural Context: La Maladie de Bradford in Local, National and Global Settings, 1878–1919' (PhD thesis, University of Leeds, 2011).

—, 'Bacteriology in the Service of Sanitation: The Factory Environment and the Regulation of Industrial Anthrax in Late-Victorian Britain', *Social History of Medicine*, 25 (2012), pp. 343–61.

—, *The Making of Modern Anthrax, 1875–1920: Uniting Local, National and Global Histories of Disease* (London: Pickering & Chatto, 2013).

Steedman, C., *Dust* (Manchester: Manchester University Press, 2000).

Stevenson, L. G., 'Science Down the Drain: On the Hostility of Certain Sanitarians to Animal Experimentation, Bacteriology and Immunology', *Bulletin of the History of Medicine*, 29 (1955), pp. 1–26.

—, 'Exemplary Disease: The Typhoid Pattern', *Journal of the History of Medicine and Allied Sciences*, 37 (1982), pp. 159–81.

Sturdy, S., 'Looking for Trouble: Medical Science and Clinical Practice in the Historiography of Modern Medicine', *Social History of Medicine*, 24 (2011), pp. 739–57.

Swiderski, R. M., *Anthrax: A History* (Jefferson, NC, and London: McFarland & Co. 2004).

Tansey, E. M., 'The Early Scientific Career of Sir Henry Dale FRS (1875–1968)' (PhD thesis, University of London, 1990).

Taplin, E. L., *Liverpool Dockers and Seamen, 1870–1890* (Hull: University of Hull Publications, 1974).

Timmermann, C., 'Constitutional Medicine, Neoromanticism and the Politics of Antimechanism in Interwar Germany', *Bulletin of the History of Medicine*, 75 (2001), pp. 717–39.

Trail, R. R. (ed.), *Munk's Roll, Volume V: Lives of the Fellows of the Royal College of Physicians of London, continued to 1965* (London: Royal College of Physicians, 1968).

Taylor, V. and F. Trentmann, 'Liquid Politics: Water and the Politics of Everyday Life in the Modern City', *Past and Present*, 211 (2011), pp. 199–241.

Tomes, N. J., 'The Private Side of Public Health: Sanitary Science, Domestic Hygiene, and the Germ Theory, 1870–1900', *Bulletin of the History of Medicine*, 64 (1990), pp. 509–39.

—, 'American Attitudes toward the Germ Theory of Disease: Phyllis Allen Richmond Revisited', *Journal of the History of Medicine and Allied Sciences*, 52 (1997), pp. 17–50.

—, *The Gospel of Germs: Men, Women, and the Microbe in American Life* (Cambridge, MA, and London: Harvard University Press, 2002).

Valenze, D., *Milk: A Local and Global History* (New Haven, CT: Yale University Press, 2011).

Vernon, K., 'Pus, Sewage, Beer and Milk: Microbiology in Britain, 1870–1940', *History of Science*, 28 (1990), pp. 289–325.

Waddington, I., *The Medical Profession in the Industrial Revolution* (Dublin: Gill and Macmillan Ltd, 1984).

Waddington, K., *Charity and the London Hospitals, 1850–1898* (Woodbridge: Boydell and Brewer, 2000).

—, *Medical Education at St. Bartholomew's Hospital 1123–1995* (Woodbridge: The Boydell Press, 2003).

—, *The Bovine Scourge: Meat, Tuberculosis and Public Health, 1850–1914* (Woodbridge: Boydell and Brewer, 2006).

Wald, P., *Contagious: Cultures, Carriers, and the Outbreak Narrative* (Durham, NC: Duke University Press, 2008).

Wall, R., 'Using Bacteriology in Elite Hospital Practice: London and Cambridge, 1880–1920', *Social History of Medicine*, 24 (2011), pp. 776–95.

Warner, J. H., 'Ideals of Science and their Discontents in Late Nineteenth-Century American Medicine', *Isis*, 82 (1991), pp. 454–78.

—, *The Therapeutic Perspective: Medical Practice, Knowledge, and Identity in America, 1820–1885* (Princeton, NJ: Princeton University Press, 1997).

—, 'The Fall and Rise of Professional Mystery: Epistemology, Authority and the Emergence of Laboratory Medicine in Nineteenth-Century America', in A. Cunningham and P. Williams (eds), *The Laboratory Revolution in Medicine* (Cambridge: Cambridge University Press, 2002), pp. 110–41.

Warwick, A., *Masters of Theory: Cambridge and the Rise of Mathematical Physics* (Chicago, IL, and London: The University of Chicago Press, 2003).

Weatherall, M. W., *Gentlemen, Scientists and Doctors: Medicine at Cambridge, 1880–1940* (Woodbridge and Rochester, NY: The Boydell Press in association with Cambridge University Library, 2000).

Webster, C., *The National Health Service: A Political History* (Oxford: Oxford University Press, 2002).

Weindling, P., 'From Medical Research to Clinical Practice: Serum Therapy for Diphtheria in the 1890s', in J. V. Pickstone (ed.), *Medical Innovations in Historical Perspective* (New York: St. Martin's Press, 1992), pp. 72–83.

White Franklin, A., 'Medical Achievements of the Eighteenth and Nineteenth Centuries', in V. C. Medvei and J. L. Thornton (eds), *Royal Hospital of Saint Bartholomew 1123–1973* (London: Saint Bartholomew's Hospital, 1974), pp. 126–84.

Whitteridge, G. and V. Stokes, *A Brief History of the Hospital of Saint Bartholomew* (London: The Governors of the Hospital of Saint Bartholomew, 1961).

Wiener, M. J., *English Culture and the Decline of the Industrial Spirit, 1850–1980* (Cambridge, Cambridge University Press, 1981).

Wilkinson, L., 'Anthrax', in K. F. Kiple (ed.), *The Cambridge World History of Human Disease* (Cambridge: Cambridge University Press, 1993), pp. 582–4.

Wilson, D., *Dockers: The Impact of Industrial Change* (London: Fontana, 1972).

Witt, J. F., *Lessons from History: State Constitutions, American Tort Law, and the Medical Malpractice Crisis*, The Project on Medical Liability in Pennsylvania funded by the Pew Charitable Trust, 2004, at http://www.pewtrusts.org/uploadedFiles/wwwpewtrustsorg/Reports/Medical_liability/medical_malpractice_witt_030904.pdf [accessed 18 May 2012].

Wohl, A., *Endangered Lives: Public Health in Victorian Britain* (London: Methuen and Co., 1984).

Wolffe, J., 'Religion and Secularization', in P. Johnson (ed.), *Twentieth-Century Britain: Economic, Social and Cultural Change* (London and New York: Longman, 1994), pp. 427–41.

Woods, A., 'Foot and Mouth Disease in 20th-Century Britain: Science, Politics and the Veterinary Profession' (PhD dissertation, University of Manchester, 2002).

Woods, R., 'Mortality and Sanitary Conditions in Late Nineteenth-Century Birmingham', in R. Woods and J. Woodward (eds), *Urban Disease and Mortality in Nineteenth Century England* (London and New York: Batsford Academic and Educational, and St. Martin's Press, 1984), pp. 176–202.

Worboys, M., *Spreading Germs: Disease Theories and Medical Practice in Britain, 1865–1900* (Cambridge: Cambridge University Press, 2000).

—, 'Was there a Bacteriological Revolution in Late Nineteenth-Century Medicine?', *Studies in the History and Philosophy of Biological and Biomedical Sciences*, 38 (2007), pp. 20–42.

Wynne, B., 'May the Sheep Safely Graze? A Reflexive View of the Expert–Lay Knowledge Divide', in S. Lash, B. Szerszynski and B. Wynne (eds), *Risk, Environment and Modernity: Towards a New Ecology* (London: Sage, 1996), pp. 44–83.

INDEX